# Waste Management in Climate Change and Sustainability Perspectives

This handbook discusses the relationships and effects of climate change on waste treatment and its sustainable management. The waste management sector is in a unique position to transition from a minor source of global greenhouse gas (GHG) emissions to a major contributor to reducing GHG emissions. This book compiles the potential impacts and benefits of various waste management systems in terms of climate impact. It investigates the global climate impact of municipal solid waste, commercial and industrial waste, agricultural waste, and hazardous waste management systems.

**Key features:**

- Reviews advanced and innovative processes for sustainable waste management
- Covers green waste treatment technologies using microbes, green soldier flies, earthworms, and bacteriophages
- Discusses the negative and positive effects of waste treatment and disposal
- Provides relevant case studies from different regions of the world
- Examines the role of climate change on emerging pollutants

This book is meant for researchers and professionals in environmental sciences, and chemical and biochemical engineering.

## Sustainable Industrial and Environmental Bioprocesses

This book series aims to provide a comprehensive collection of books focusing on bioprocesses in industrial and environmental biotechnology. The multidisciplinary content encompasses chemical and biochemical engineering, industrial microbiology, and energy biosciences, all with a central theme of sustainable development and circular economy principles. The books align with the Sustainable Development Goals (SDGs) and offer state-of-the-art information and in-depth knowledge on the subject matter. The books in this series also emphasize on the application of emerging tools, such as machine learning and artificial intelligence, for the advancement of bioprocesses. While primarily targeting academicians and researchers, this series is valuable for policy planners and industry professionals, with carefully tailored contents to cater to their specific needs and interests.

*Series Editor: Dr Ashok Pandey*

**Professor Ashok Pandey** is currently a Distinguished Scientist at the Centre for Innovation and Translational Research, CSIR-Indian Institute of Toxicology Research, Lucknow, India. His major research and technological development interests are industrial and environmental biotechnology and energy biosciences, focusing on biomass to biofuels and chemicals, waste to wealth and energy, etc. He has 16 patents, 120 books, more than 1000 papers and book chapters, etc., with an h-index of 131 and more than 74,000 citations. Professor Pandey is the recipient of many national and international awards and honors, which include Fellow, The World Academy of Sciences (TWAS); Highly Cited Researcher (top 1% in the world), Clarivate Analytics (since 2018 till date); Rank 1 in India in Biology and Biochemistry and Rank 417 in the world, Research.Com (2023); Rank 1 in India in Microbiology under Enabling and Strategic Technologies Sector, Elsevier (2021); and Rank 1 in India in Biotechnology and Rank 8 in the world, Stanford University Report (2020–2021–2022), etc.

*Pharmaceuticals in Aquatic Environment: Remediation Technologies and Future Challenges*
Vinod Kumar Garg, Ashok Pandey, Navish Kataria, Caterina Faggio

*Biomass Hydrolyzing Enzymes: Basics, Advancements, and Applications*
Reeta Rani Singhania, Anil Kumar Patel, Héctor A. Ruiz and Ashok Pandey

*Decentralized Sanitation and Water Treatment: Concept and Technologies*
R.D. Tyagi, Ashok Pandey, Patrick Drogui, Bhoomika Yadav, Sridhar Pilli and Jonathan W.C. Wong

*Decentralized Sanitation and Water Treatment: Treatment in Cold Environments and Techno-Economic Aspects*
R.D. Tyagi, Ashok Pandey, Patrick Drogui, Bhoomika Yadav, Sridhar Pilli and Jonathan W.C. Wong

*Waste Management in Climate Change and Sustainability Perspectives: Organic and Medical Waste*
Sunita Varjani, Izharul Haq, Ashok Pandey, Vijai Kumar Gupta, Xuan-Thanh Bui

For more information about this series, please visit: https://www.routledge.com/Sustainable-Industrial-and-Environmental-Bioprocesses/book-series/SIEB

# Waste Management in Climate Change and Sustainability Perspectives

## Organic and Medical Waste

Edited by

Sunita Varjani
City University of Hong Kong, Hong Kong

Izharul Haq
Dr. B. Lal Institute of Biotechnology, Jaipur, India

Ashok Pandey
CSIR-Indian Institute of Toxicology Research, Lucknow, India

Vijai Kumar Gupta
Dublin City University, Ireland

Xuan-Thanh Bui
Ho Chi Minh City University of Technology,
Vietnam National University Ho Chi Minh, Vietnam

CRC Press
Taylor & Francis Group
Boca Raton London New York

CRC Press is an imprint of the
Taylor & Francis Group, an **informa** business

Designed cover image: Shutterstock

First edition published 2024
by CRC Press
2385 NW Executive Center Drive, Suite 320, Boca Raton FL 33431

and by CRC Press
4 Park Square, Milton Park, Abingdon, Oxon, OX14 4RN

CRC Press is an imprint of Taylor & Francis Group, LLC



ISBN: 9781032439075 (hbk)
ISBN: 9781032480046 (pbk)
ISBN: 9781003386902 (ebk)

DOI: 10.1201/9781003386902

Typeset in Times
by codeMantra

# Contents

## SECTION C: CASE STUDIES

# Preface

This book *Waste Management in Climate Change and Sustainability Perspectives: Organic and Medical Waste* aims to address various aspects of organic and medical waste treatment and management, including the relationship of waste management with climate change, case studies related to waste management, and cutting-edge treatment technologies. The waste management sector has the potential to transition from being a minor source of global greenhouse gas (GHG) emissions to playing a significant role in reducing such emissions. While waste treatment and disposal release some emissions, preventing and recovering value-added products from waste can avoid emissions. Waste management practice generates GHGs both directly and indirectly. By adopting a comprehensive approach to waste management, it is possible to positively impact GHG emissions across various sectors such as agriculture, energy, and manufacturing. In response to climate change, numerous waste management programs have been developed, including Integrated Solid Waste Management (ISWM) based on the 5R (Refuse, Reduce, Reuse, Repurpose, and Recycle) principles, the conversion of waste biomass into energy or value-added products.

The primary objective of this book is to examine how diverse waste management practices can contribute to climate change mitigation. It aims to investigate the relationship between waste and climate change. This book also intends to evaluate the potential advantages and consequences of various waste management technologies concerning climate change, utilizing existing literature as a basis. It offers a comprehensive global analysis of the climate impact associated with the management of different waste types, encompassing municipal solid waste (MSW), industrial waste and biomedical waste, etc. Chapters of this book offer detailed and updated information on waste management, climate change, and related areas. This book serves as a comprehensive guide for students, scientists, and academicians working in environmental biotechnology and engineering, including relevant areas of waste management.

This book comprises ten chapters in three sections. Section A, "Waste Management and Climate Change," delves into the latest advancements in addressing organic waste management. Chapter 1 highlights emerging green technologies, such as anaerobic digestion and vermicomposting, which offer promising solutions for sustainable waste management practices. Chapter 2 explores the role of phages in natural and directed microbial control, presenting a unique avenue for disease prevention in organic waste management. Chapter 3 delves into biodegradable waste management technologies, emphasizing their potential to mitigate climate change and reduce GHG emissions. In addition, Chapter 4 focuses on rehabilitating ecosystem services in the face of challenging climate change conditions.

Section B, "Waste Management and Sustainability," examines the intersection of waste management and the circular economy paradigm. Chapter 5 explores sustainable waste management practices and their alignment with the principles of the circular economy, promoting economic efficiency and environmental conservation. Chapter 6 presents waste as a valuable resource for the production of value-added products, advocating for the integration of recycling and upcycling techniques within the circular economy framework. Furthermore, Chapter 7 discusses biodrying of municipal solid wastes for the production of Refused Derived Fuel (RDF), showcasing its potential as a sustainable waste management solution. In Section C, "Case Studies," real-world examples showcase the practical application of waste management strategies. Chapter 8 presents a case study on large-scale rotary drum composting, demonstrating its effectiveness in urban organic waste management. Chapter 9 highlights the integration of technologies for value-added products from industrial waste, focusing on the context of Ho Chi Minh City. Lastly, Chapter 10 addresses the critical issue of biomedical waste management during the COVID-19 pandemic, providing insights into effective waste disposal practices.

We would like to express our heartfelt appreciation to the authors for their contribution based on their expertise and knowledge. Their dedication and commitment to addressing the critical issues of waste management, climate change, and sustainability have been instrumental in shaping the content of this publication. We are grateful to the reviewers who generously shared their valuable comments, suggestions, and scientific insights. Their careful evaluation and constructive feedback played a crucial role in refining the chapters and enhancing the scientific discussions presented in this book.

We sincerely thank the CRC Press team comprising Dr. Bhavik Sawhney, Senior Commissioning Editor; Ms. Neha Bhatt, Editorial Assistant; and the entire CRC Press production team for their support in publishing this book. Their expertise, professionalism, and support throughout the publication process have been invaluable in bringing this book to fruition.

*Editors*
**Sunita Varjani,** Hong Kong
**Izharul Haq,** India
**Ashok Pandey,** India
**Vijai Kumar Gupta,** Ireland
**Xuan-Thanh Bui,** Vietnam

# About the Editors

**Dr. Sunita Varjani** is currently working as a visiting fellow at the City University of Hong Kong, Hong Kong. Her major areas of research are Industrial and Environmental Biotechnology, Wastewater Treatment and Process Engineering, and Bioprocess Technology and Waste Management. She has worked as a visiting scientist at EPFL, Lausanne, Switzerland. Dr. Varjani has authored more than 400 publications, including research and review papers, books, book chapters, and conference communications with h index of 67 and > 17,800 citations.

**Dr. Izharul Haq** an Assistant Professor at Dr. B. Lal Institute of Biotechnology, Jaipur, India. He had worked as a Post-Doctoral Fellow in the Indian Institute of Technology Guwahati, India. He obtained his PhD in Microbiology from CSIR-Indian Institute of Toxicology Research, Lucknow, India. He is working on liquid and solid waste management through microorganisms and their toxicity evaluation. Dr. Haq has 62 publications with h index of 18 and > 1400 citations.

**Prof. Ashok Pandey** is currently Distinguished Scientist at the Centre for Innovation and Translational Research, CSIR-Indian Institute of Toxicology Research, Lucknow, India. His major research and technological development interests are industrial & environmental biotechnology and energy biosciences, focusing on biomass to biofuels & chemicals, waste to wealth & energy, etc. He has 16 patents, 120 books, >1000 papers and book chapters, etc with *h* index of 139 and >81,400 citations.

**Dr. Vijai Kumar Gupta** obtained a doctoral degree in Microbiology from Dr RML Avadh University, India. Currently, he is working as an Associate Professor in Dublin City University, Ireland. His focus of research is on developing bioprocess technologies to valorize available bioresources for the production of value-added biochemicals, bioproducts, and fuels and their potential to address the key challenges in agri-food-pharma-environment importance. He has >500 publications, including 48 book chapters and 42 books with h-index 81 and > 28,000 citations.

**Professor Xuan-Thanh Bui** obtained his PhD degree in environmental engineering from Asian Institute of Technology, Thailand. He currently serves as the chairman of the Department of Water Science & Technology and the head of Key Laboratory of Advanced Waste Treatment Technology in Ho Chi Minh City University of Technology. His research focuses on membrane processes, water and waste treatment technologies, and green technologies. He has published >250 papers ($6{,}800^{+}$ citations, h-index=44), eight books, 36 book chapters, and has four patents with h index of 44 and >6800 citations.

# Contributors

**Amro Abd Al Fattah Amara**
Protein Research Department
Genetic Engineering and Biotechnology Research Institute
Universities and Research Centre District
Alexandria, Egypt

**Anouar Bardi**
Higher Institute of Management of Gabès
Gabès, Tunisia

**Suzana Fátima Bazoti**
Laboratory of Microbiology and Bioprocesses (LAMIBI)
Federal University of Fronteira Sul
Erechim, Brazil
and
Department of Chemical and Food Engineering
Federal University of Santa Catarina
Santa Catarina, Brazil

**Xuan-Thanh Bui**
Key Laboratory of Advanced Waste Treatment Technology & Faculty of Environment and Natural Resources
Ho Chi Minh City University of Technology (HCMUT)
Ho Chi Minh City, Vietnam
Vietnam National University Ho Chi Minh City (VNU-HCM)
Ho Chi Minh City, Vietnam

**Chart Chiemchaisri**
Department of Environmental Engineering
Faculty of Engineering
Kasetsart University
Chatuchak, Bangkok, Thailand

**Wilai Chiemchaisri**
Department of Environmental Engineering, Faculty of Engineering
Kasetsart University
Chatuchak, Bangkok, Thailand

**Caroline Dalastra**
Laboratory of Microbiology and Bioprocesses (LAMIBI)
Federal University of Fronteira Sul
Erechim, Brazil

**Chaichi Devi**
JAIN (Deemed-to-be-University)
Bengaluru, India

**Jean Pierre Doussoulin**
Instituto de Economía
Universidad Austral de Chile
Los Rios, Chile
Université Gustave Eiffel
Research Group on the Use of Panel Data in Economics (Erudite)
Créteil, France

**Jeanger P. Juanga-Labayen**
Carlos Hilado Memorial State University, Negros Occidental
Negros Occidental, Philippines

**Ho Thi Ngoc Ha**
Key Laboratory of Advanced Waste Treatment Technology & Faculty of Environment and Natural Resources
Ho Chi Minh City University of Technology (HCMUT)
Ho Chi Minh City, Vietnam
and
Vietnam National University Ho Chi Minh City (VNU-HCM)
Ho Chi Minh City, Vietnam

**Ho Thi Thanh Hien**
Faculty of Environment
School of Engineering and Technology
Van Lang University
Hồ Chí Minh, Vietnam

**Qahraman Kakar**
Université Gustave Eiffel
Research Group on the Use of Panel Data in Economics (Erudite)
Créteil, France

**Ajay S. Kalamdhad**
Department of Civil Engineering
Indian Institute of Technology Guwahati
Guwahati, India

**Nguyen Dang Khoa**
Faculty of Environment
School of Engineering and Technology
Van Lang University
Ho Chi Minh, Vietnam

**Meena Khwairakpam**
School of Agro and Rural Technology
Indian Institute of Technology Guwahati
Guwahati, India

**Gabriel Henrique Klein**
Laboratory of Microbiology and Bioprocesses (LAMIBI)
Federal University of Fronteira Sul
Erechim, Brazil

**Simone Kubeneck**
Laboratory of Microbiology and Bioprocesses (LAMIBI)
Federal University of Fronteira Sul
Erechim, Brazil
Faculty of Environment
School of Engineering and Technology
Van Lang University
Hồ Chí Minh, Vietnam

**Ildefonso Villavecer Labayen**
Technological University of the Philippines Visayas
Negros Occidental, Philippines

**Linh-Thy Le**
Faculty of Public Health
University of Medicine and Pharmacy at Ho Chi Minh City (UMP)
Ho Chi Minh City, Vietnam

**Huynh Tan Loi**
Faculty of Environment
School of Engineering and Technology
Van Lang University
Hồ Chí Minh, Vietnam

**Phuong Vu Mai Doan**
Rowan University
Nagasaki University
Nagasaki, Japan

**Melissa Montalbo-Lomboy**
Rowan University
Glassboro, New Jersey, USA

**Júlia Nerling**
Laboratory of Microbiology and Bioprocesses (LAMIBI)
Federal University of Fronteira Sul
Erechim, Brazil

**My Thi Tra Ngo**
Ho Chi Minh City University of Technology
Ho Chi Minh, Vietnam

**Huu Hao Ngo**
Centre for Technology in Water and Wastewater
School of Civil and Environmental Engineering
University of Technology
Sydney, Australia

**Le Thi Kim Oanh**
Faculty of Environment
School of Engineering and Technology
Van Lang University
Hồ Chí Minh, Vietnam

**Panida Payomthip**
The Joint Graduate School of Energy and Environment (JGSEE)
King Mongkut's University of Technology Thonburi
Thonburi, Bangkok, Thailand

**Ruchira Perera**
Department of Environmental Engineering
Faculty of Engineering
Kasetsart University
Chatuchak, Bangkok, Thailand

**Suryateja Pottipati**
Department of Civil Engineering
Indian Institute of Technology Guwahati
Guwahati, India

**Hayfa Rajhi**
University Hospital of Gabès
Tunisia Health Ministry
Gabès, Tunisia

**Noppharit Sutthasil**
Department of Environmental Engineering
Faculty of Engineering
Kasetsart University
Chatuchak, Bangkok, Thailand

**Sakulrat Sutthiprapa**
Department of Environmental Engineering
Faculty of Engineering
Kasetsart University
Chatuchak, Bangkok, Thailand

**Sirintornthep Towprayoon**
The Joint Graduate School of Energy and Environment (JGSEE)
King Mongkut's University of Technology Thonburi
Thonburi, Bangkok, Thailand

**Helen Treichel**
Laboratory of Microbiology and Bioprocesses (LAMIBI)
Federal University of Fronteira Sul
Erechim, Brazil

**Sunita Varjani**
School of Energy and Environment
City University of Hong Kong
Hong Kong, Hong Kong
and
Sustainability Cluster, School of Engineering
University of Petroleum and Energy Studies
Dehradun, India

**Komsilp Wangyao**
The Joint Graduate School of Energy and Environment (JGSEE)
King Mongkut's University of Technology Thonburi
Thonburi, Bangkok, Thailand

# *Section A*

## *Waste Management and Climate Change*

# 1 Emerging Green Technologies for Organic Waste Management

*Jeanger P. Juanga-Labayen,*
*Ildefonso Villavecer Labayen, Melissa Montalbo-Lomboy,*
*My Thi Tra Ngo, and Phuong Vu Mai Doan*

## 1.1 INTRODUCTION

Waste has been of paramount importance for human existence since the pre-historic period. Waste generation can be attributed to man's daily activities. Technological and economic development, which catalyzed the growth of the population and urbanization, along with the changes in human lifestyle all contributed to the complexity of the characteristics of waste and accelerated waste generation. According to Kaza et al. (2018), global waste generation will significantly surpass population growth by more than twice by 2050. The forecasted global annual waste generation of 2.59 billion tons by 2030 is projected to reach 3.40 billion tons by 2050. Furthermore, more than 50% of the global waste composition consists of organic waste, and a large fraction of this waste is disposed in open dumps and unspecified landfills. Despite global progress in solid waste management, there is still an apparent need to integrate waste management into the paradigm of economic growth and technological innovation along with the pressing dilemma of climate change. Solid waste management affects everyone, especially in cities and countries that grow rapidly, which desperately need systems to manage their burgeoning waste. Hence, holistic planning and managing of resources along with innovative technologies are urgently needed.

Green technology can be defined as the application of science and technology in transforming waste materials into a valuable resource that supports environmental sustainability to reduce human-induced pollution. There were several articles on green technology for different areas of environmental concern; however, a comprehensive study on emerging green technology for organic waste treatment was found limited. Hence, this chapter covers the important emerging green technologies such as fermentation, anaerobic digestion, vermicomposting, and bioremediation in transforming organic waste into sustainable products such as bioplastics, biofuels, and biofertilizers. These technologies utilize microorganisms that convert organic matter into valuable resources; however, they may differ in the process parameters.

The fermentation process can produce biodegradable plastics via enzymatic reaction and the use of *Lactobacillus* genus. Anaerobic digestion consists of a series of microbial interactions under anoxic conditions to generate methane-rich fuel. On the other hand, vermicomposting involves the bioconversion of organic waste into peat-like, nutrient-rich humus through the interactions of microorganisms and earthworms. Bioremediation offers a sustainable solution to transform complicated pollutants into a safe or nontoxic form of substrate that has the potential to be used as a nutrient-rich substance.

DOI: 10.1201/9781003386902-2

## 1.2 FERMENTATION

Fermentation is a process that involves microorganisms converting organic materials into potentially useful and high-value materials. It is widely known that fermentation is the method used to produce many foods and beverages. However, it is essential to note that the application of fermentation has evolved over time. In 2016, the world's earliest evidence of fermentation was found in Norje Sunnansund, situated in the southeastern part of Sweden (Boethius, 2016). The archaeologist found evidence that large quantities of fish were fermented without salt for food preservation more than 9000 years ago (Boethius, 2016). Beer brewing has been associated with fermentation since its invention. In 2018, another group of archaeologists reported a 13,000-year-old stone mortar in a Natufian graveyard at the Raqefet Cave in Israel (Liu et al., 2018). After thorough analysis, they found that the Natufians used the stone mortar to grind cereal-based plants such as wheat and barley to make beer as part of their ritual to honor their dead. These findings indicate that fermentation existed as far as 13,000 years ago or even further. What started as a way to produce food and beverages in ancient times has evolved into a quintessence of modern civilization. As of today, fermentation technology has boundless applications in food, beverage, plastics, biochemicals, pharmaceuticals, cosmetics, detergents, animal feed, and wastewater treatment.

Every living organism on Earth has basic needs to survive. For example, humans will require food, water, air, and shelter to survive. A microorganism that undergoes fermentation would need some form of energy source to survive. Based on the microorganism's nutritional requirements, they are categorized as follows: photoautotrophs, photoheterotrophs, chemoautotrophs, and chemoheterotrophs. Photoautotrophs and photoheterotrophs will require light as energy, while chemoautotrophs and chemoheterotrophs will use chemicals as energy sources (Kampen, 2014). Photoautotrophs and chemoautotrophs utilize $CO_2$ as a carbon source, while photoheterotrophs and chemoheterotrophs use organic compounds as a carbon source. These microorganisms, mainly bacteria, yeast, and fungi, will require carbon, hydrogen, oxygen, nitrogen, sulfur, and other trace elements to grow and biosynthesize. In this chapter, the following sections that discuss fermentation will focus on utilizing organic waste to produce valuable and high-value products such as bioplastics.

### 1.2.1 Bioplastics via Fermentation

Synthetic plastics have been part of our lives since the discovery of Bakelite in 1907 (Crespy et al., 2008). Plastics, as we know them today, are synthetic polymers molded into various shapes and sizes. They are present in almost everything we use. Plastic is used in our homes, workplaces, hospitals, schools, and even space. It became prevalent because of its very versatile properties, such as being lightweight with varying degrees of strength; thermal and electrical insulators, resistant to chemicals, can be processed into a limitless range of shapes, colors, and sizes; and most of all, it is very cheap compared to other materials, notably metals and glass. However, one of the significant flaws of synthetic plastics is their inability to biodegrade when discarded. According to ourworldindata.org (Ritchie & Roser, 2018), the global annual plastic production has reached 460 million tons in 2019 and generates 302 million tons as waste. In 2015 (Ritchie & Roser, 2018), it was reported that 25% of plastic waste is incinerated and 20% is recycled. Considering that incineration and recycling were almost nonexistent in the 1980s, this is already a significant achievement. In addition, the production of synthetic plastics has emitted massive greenhouse gases, resulting in climate change.

One of the solutions to the plastic problem is the utilization of bioplastic. "Bioplastics are bio-based, biodegradable plastics or both" (Bioplastics, 2021). It should be noted that bio-based and biodegradable plastics are not the same. Bio-based products are defined "as a commercial or industrial (other than food or feed) that is composed, in whole, or in significant part, of biological products, including renewable domestic agricultural materials, renewable chemicals, and forestry materials or an intermediate ingredient or feedstock" (USDA, 2023). Biodegradable plastics are those that biologically break down when exposed to microorganisms. Bio-based, or partially bio-based, plastics can be nonbiodegradable plastics such as bio-based polypropylene (PP) or bio-based polyethylene terephthalate (PET), or can be biodegradable such as polylactic acid (PLA) or polyhydroxyalkanoate (PHA). Some plastics are petroleum-based but biodegradable, such as polybutylene adipate terephthalate (PBAT).

PLA and PHA have been gaining commercial interest because of their comparable strength to synthetic polymers, biodegradability, and biocompatibility properties. According to European Bioplastics (Bioplastics, 2022), the current global production capacities of bioplastic, including PLA, PHA, starch, and others, are estimated to be 2.23 million tons and are expected to increase to 6.3 million tons by 2027. Currently, PLA is produced through the fermentation of carbohydrates such as cornstarch and sugarcane. The *Lactobacillus* genus, such as *L. delbrueckii*, *L. atmophiles*, and *L. bulgaricus*, consumes the carbohydrate and produces lactic acid or lactide monomer that is the building block for PLA. On the other hand, PHA can be made via enzymatic reaction, transgenic plants, or microbial fermentation. Among the three methods, the microbial fermentation is the most extensively studied.

The successful commercialization of PLA and PHA depends on several factors including fermentation feedstock, product yield, production rate, and downstream processing. PLA and PHA are industrially produced via fermentation using virgin raw materials such as cornstarch and sugarcane for PLA and sugars and plant oils for PHA. With the current industrial technologies, bioplastic is more expensive than petroleum-based plastic. The average cost of petroleum-based plastic is estimated to be $1,000–1,500 per MT compared to bioplastics at $4,000–15,000 per MT (Kosseva & Edy, 2018; Talan et al., 2020). In a 2022 review (Rosenboom et al., 2022), PET and polystyrene (PS) cost $1.2–1.4 and $0.7–1.5 per kg, respectively, while PLA cost $2–3 per kg and poly-3-hydroxybutyrate (P3HB) cost $3–8 per kg. Owing to the high cost of virgin raw materials and the increasing demand to food, utilizing organic waste streams for bioplastic becomes an attractive solution.

### 1.2.2 Organic Waste Streams for Polyhydroxyalkanoate (PHA) and Polylactic Acid (PLA)

The fermentation process allows conversion of waste materials to valuable products like bioplastic. The utilization of organic waste streams for PHA production has been extensively studied (Rajendran & Jeehoon, 2022; Bhatia et al., 2021; Saratale et al., 2021; Yadav et al., 2020). One of the most considerable resources for PHA is the lipid-rich waste from dairy farms, edible oil plants, animal fats, restaurants, and household cooking oil. A review report has shown that lipid feedstock has a higher PHA yield and costs significantly less than conventional feedstock (Bhatia et al., 2021). In addition, with the surge in biodiesel production worldwide, there is an enormous surplus of its by-product, glycerol. Several studies have also investigated glycerol as a viable feedstock for PHB production (Fauzi et al., 2019; Burniol-Figols et al., 2018; Zhu et al., 2013). PHB or polyhydroxybutyrate is one of the most common types of PHA. In terms of carbon source, high organic waste or byproducts such as sugarcane molasses (Acosta-Cárdenas et al., 2018), cheese whey, olive oil and winery wastewater (Gottardo et al., 2022), and even municipal wastewaters (Mannina et al., 2020) have been explored for PHA production. Even with the promise of the cost-effectiveness of organic waste as feedstock in PHA production, more research still needs to be done before its commercialization. The downstream processing of PHA production is still very complex and costly, to compete with petroleum-based plastics.

PLA is considered a bio-based, recyclable, and compostable polymer or plastic created through the fermentation of renewable materials like cornstarch, sugarcane, or sugar beet. This fermentation process produces lactic acid monomers which, when combined, forms PLA. The banana peels, coffee grounds, rotten tomatoes, moldy bread, and other food waste that ends up in the landfill could be transformed into a biodegradable plastic packaging material creating an alternative to single-use plastics (FoodPrint, 2019). PLA production can have a high impact compared with petrochemical plastic production, but diversion of PLA and food waste to be organically recycled, via anaerobic digestion, or potentially insect feed in the future, can improve the overall environmental performance of the bioplastic packaging (Bishop et al., 2021). PLA is often marketed as the biodegradable or compostable bio-based plastic alternative to petroleum-based plastics that can be used for food packaging (TotalEnergies, 2023; Bishop et al., 2021).

## 1.3 ANAEROBIC DIGESTION

Anaerobic digestion is a widespread technology applied in various waste streams including industrial and municipal wastewater, organic fractions of municipal solid waste, food waste, and agricultural waste. Co-digestion of different substrates by mixing two or more waste streams offers better process capability in treating heterogeneous waste streams. Anaerobic digestion of organic waste was the preferred approach and reliable technology for the provision of energy and reduction of greenhouse gas emissions when compared to combustion/incineration, aerobic composting, pyrolysis, and landfilling/landfill gas recovery.

Anaerobic digestion also known as biomethanation or biomethanization is a biochemical degradation process that converts the complex organic substrate into simpler constituents in a series of metabolic interactions of a wide spectrum of microorganisms that catalyze the process under anoxic conditions. In simple terms, anaerobic digestion is a natural process of biomass degradation in the absence of oxygen. It is a versatile process that is capable of capturing the biogas potential of organic waste through a natural fermentation process. This process supports the generation of renewable energy in the form of biogas and a nitrogen-rich digestate or stable humus from digested organic waste otherwise referred to as effluent, thus reducing methane emissions. The digestate produced from the anaerobic digestion of organic waste is a mixture of bacterial biomass and inert organics. The digestate primarily depends on the quality of the feedstock or substrate, bacterial community, the operating parameters employed, the design of an anaerobic digestion system, and digestate processing techniques. Dewatering of the digestate is typically not necessary for a dry anaerobic digestion system (Logan & Visvanathan, 2019).

Typically, between 40% and 60% of the organic matter present in the feedstock is converted to biogas. Raw biogas mainly consists of methane and carbon dioxide. It also contains trace levels of water, hydrogen sulfide, ammonia, hydrogen, and various volatile organic compounds (VOCs) depending on the feedstock quality (Wellinger et al., 2013). Table 1.1 presents the typical biogas composition and its concentration from biowaste. Biogas with a methane content higher than 45% is flammable; the higher the methane content, the higher the energy value of the gas (Deublein & Steinhauser, 2011).

Generally, the anaerobic digestion of organic waste involves three main phases that involve pretreatment, anaerobic digestion process, and posttreatment. Pretreatment is the first phase of the overall anaerobic digestion system, which aims to increase the biodegradability of the substrate by exposing the nutrients to most microbial species, thereby enhancing process efficiency and increasing the biogas yield. Pretreatment mainly involves the segregation of inert, impurities, and bulky materials from the organic waste. It involves size reduction that provides a smaller particle size of feedstock for efficient digestion, mixing the waste with other substrates into the desired consistency, and pressing the feedstock. Pretreatment methods involve mechanical, thermal, chemical,

**TABLE 1.1**
**Typical Composition and Concentration of Biogas**

| | Agricultural | | Waste | | Landfill | |
|---|---|---|---|---|---|---|
| Components (Vol %) | Range | Mean | Range | Mean | Range | Mean |
| $CH_4$ | 49–69 | 56 | 44–67 | 56 | 40–70 | 53 |
| $CO_2$ | 29–44 | 40 | 30–44 | 37 | 25–40 | 35 |
| $N_2$ | 0.6–13 | 3 | 0.1–6 | 1 | 0–17 | <5 |
| $O_2$ | 0.2–3 | 0.6 | 0.1–3 | 0.6 | 0–3 | <1 |

*Source:* Adapted from Calbry-Muzyka et al. (2022).

and enzymatic processes that convey the partial or complete decomposition of a substrate to form fermentable organics and reduce the lignin content or crystallinity of cellulose. Advancements in pretreatment technologies overcome the structural barriers of hardly biodegradable lignocellulose and its polymers (cellulose and hemicellulose) from microbial decomposition to capture the biogas potential of feedstock.

A typical anaerobic digestion facility utilized pre-treated biowaste as feedstock to the digester to produce biogas and fertilizer. Biogas can be used in engines for combined heat and power and electricity burned to produce heat or can be cleaned and used as natural gas or as vehicle fuel. While the stable digestate in a solid form similar to peat is used as a soil conditioner, the liquid form is used as fertilizer that promotes the return of nutrients to the soil. Anaerobic digestion offers significant benefits that include recycling nutrients back to the land in the form of digestate or fertilizer; reduction of greenhouse gas emissions; conserving landfill space by diverting biodegradable waste from landfill; generation of renewable energy; creation of jobs in the supply chain; heat capture and use in heating schemes; and upgrading biogas to be used as a biofuel for vehicles and to natural gas quality for injection into the gas grid (US EPA, 2016, 2022). Hence, this technology is considered one of the earliest approaches to a circular bioeconomy. Furthermore, anaerobic digestion can produce biohythane utilizing organic waste substrate like food waste. However, at the moment, biohythane production is limited only to laboratory scale and pilot scale (Bolzonella et al., 2019).

Posttreatment processes after anaerobic digestion involve dewatering and aeration to produce a nutrient-rich stable digestate with reduced organic carbon, and the liquid part has the potential to be used as a liquid fertilizer (Wang et al., 2021). Recent studies focus on extracting reusable nutrients from digestate, and nutrient recovery from liquor promotes sustainable waste management and circular economy (Sfetsas et al., 2022). The promising results of posttreatment (biological, mechanical, thermal, chemical, and thermo-chemical) and the economic feasibility should be examined carefully (Romio et al., 2021).

### 1.3.1 Anaerobic Digestion Technologies

There are several technologies in which anaerobic digestion can be used. In terms of the substrates' total solid (TS) content, the digestion can be wet (≤10% TS), semi-dry (10–20% TS), or dry (≥20%). However, such percentages of TS vary across the literature (Kumar & Samadder, 2020). Generally, the wet system also known as low solids anaerobic digestion has TS of ≤10%, whereas the dry or high solid system contains TS of ≥20%, typically 20%–50% (Abbassi-Guendouz et al., 2012). The dry system is preferred over the wet method because it has a number of benefits including a high organic loading rate, smaller reactor volume, less energy requirement, and simpler digestate handling (Komilis et al., 2017; Jain et al., 2015). Digestate from a dry system can be pelletized into biomass fuel to make briquettes and can be utilized as a fertilizer or a soil conditioner. The dry digesting technique, however, is only suitable for the batch system. In a wet system, the waste is first turned into a pulp prior to being processed and can be performed in a single- or two-stage system. A wet system is mainly used for the treatment of animal manure, sewage, and industrial waste (Rapport et al., 2012).

In terms of operating temperature, anaerobic digestion systems can be mesophilic (30–40°C) or thermophilic (45–60°C) systems. Mesophilic systems are considered more stable and require less energy input than thermophilic digestion systems. However, thermophilic digesters are more efficient in terms of retention time, loading rate, biogas production, and sanitizing digestate. In thermophilic systems, process instability has been linked to higher volatile fatty acids production that causes a pH drop (Shi et al., 2013). Thus, the thermophilic system requires a significant level of operational control and monitoring.

In terms of substrate feeding mode, anaerobic digestion can be a batch or continuous process. Batch-fed digesters are initially filled with feedstock and left tightly closed until the digester is completely exhausted of biogas production, and then opened again and emptied (Khalid et al., 2011). It is the simplest, cheapest, and oldest technology. On the contrary, continuous digestion involves feeding new feedstock at regular intervals, while an equivalent amount of slurry leaves the digester's effluent, thereby providing a continuous digestion process.

On the basis of the number of stages, the anaerobic digestion can be single- or two-stage systems. In a single-stage digestion system, all of the biological reactions occur within a single reactor. In a multi-stage digestion system, two or more different digesters are used to optimize the reactions (DEFRA, 2011). The most popular single-stage systems that employ dry systems (≤20% TS) are Dranco, Kompogas, Valorga, and Bekon (Kumar & Samadder, 2020; Li et al., 2011). A single-stage system is used mainly for small decentralized units that are easy to design, build and operate, and less expensive than a two-stage system (Rapport et al., 2012). The two-stage system aims to sequence the biochemical reactions to optimize the process. For example, BTA (Biotechnische Abfallverwertung GmbH & Co. KG) system employs a wet, two-stage digestion process for the organic fraction of waste consisting of reactors for the hydrolysis/acidogenesis stage and another reactor for the methanogenesis stage (BTA-Biotechnical Recycling, 2022).

### 1.3.2 Feedstock for Anaerobic Digestion

"Substrate" or "feedstock" refers to biomass used for anaerobic digestion. Anaerobic digestion can treat a wide range of organic wastes. Historically, liquid wastes, such as sewage, industrial wastewater, manure, and sludge, have been treated using anaerobic digestion. In the mid-20th century, solid waste with a high organic content attracted anaerobic digestion. Table 1.2 provides an overview of the various feedstocks from municipal, agricultural, and industrial sources.

The organic fraction of municipal solid waste (OFMSW) is made up of a variety of scraps from homes, restaurants, retail stores, parks, gardens, and the food sector. It typically consists of organic wastes including food trash, yard waste, paper, and newspaper. Its composition and production are influenced by geographic location, population density, socioeconomic activity, dietary preferences,

**TABLE 1.2**
**Organic Feedstock for Anaerobic Digestion**

| Municipal | Agriculture | Industry |
|---|---|---|
| • Organic fraction of municipal solid waste | • Livestock manure | • Slaughterhouse waste |
| • Human excreta | • Energy crops | • Food processing waste |
| | • Algal biomass | • Pulp and paper waste |
| | • Agro-industrial waste | |

cultural factors, climatic factors, and collection methods. This waste stream is often distinguished by its high moisture content and biodegradable material. In most cases, anaerobic digestion reduces the solids' content by 50%–60% while preserving nutrients for the soil and eradicating up to 95% of any pathogenic organisms. The resultant digestate is a stable, low-odor, high-nutrient product that can be applied to land. This process produces biogas, which can either be processed and provided to utilities for natural gas use or utilized to produce power. A typical ton of OFMSW processed yields 100–200 $m^3$/ton of biogas (Kaza & Bhada-Tata, 2018). According to Juanga-Labayen et al. (2022), cotton is another potential substrate for anaerobic digestion as it contains more than 50% cellulose. Cotton waste (cotton stalks, cottonseed hull, and cotton oil cake), cotton waste from spinning mills, and cotton textile waste can be treated anaerobically to produce methane-rich biogas. Pretreatment is a proven method to enhance the anaerobic degradability of the substrate. Cotton substrate treated with 0.5 M $Na_2CO_3$ at 150°C for 3 h shows an increased methane yield and anaerobic degradability of more than 50% when compared to untreated substrate at substrate to inoculum ratio (SIR) of 1.0, 1.5, and 2.0. Furthermore, higher substrate loading of SIR 1.0 produced a comparable yield of 306.73 mL $CH_4$/g VS, which is equivalent to anaerobic degradability of 75% for treated substrate (Juanga-Labayen et al., 2021).

According to municipal sources, human waste or human excreta is another potential substrate for biogas production (Vögeli et al., 2014). However, when using human waste for methane production, there is a need for proper separation from urine, collection, and storage to facilitate in situ treatment. In situ digestion systems for human waste are desirable to avoid transfer and minimize the movement of pathogens as it is a pathogen-rich waste. Utilizing human waste was viewed as a highly abundant substrate and sustainable because of the growing human population. However, the presence of ammonia found in human waste inhibits the anaerobic digestion process that would produce impure biogas with high nitrogen concentrations. Emetere et al. (2022) studied the co-digestion of human excreta with powdered chicken feathers, which drastically reduces the high nitrogen content in the biogas and improves methane concentration in biogas. Utilizing chicken as an optimizing agent to enhance biogas from human excreta can potentially lead impoverished or rural areas to generate their own sustainable source of energy for cooking and lighting (Emetere et al., 2022; Corless, 2022).

Agricultural waste encompasses crop residues from farming activities and is a potential resource for biogas production. The use of anaerobic digestion technology as an effective way of agricultural waste treatment to generate biogas offers a sustainable solution to address environmental contamination and minimize carbon emissions through the use of renewable energy in the form of biogas (Lonnqvist et al., 2015). Agricultural waste residues mainly consist of straw from crops, e.g., rice, wheat, corn, soybeans, potatoes, cotton, and sugarcane and livestock manure from pigs, cattle, cows, sheep, and poultry. Liu et al. (2023) estimated that the available biogas potential of agricultural wastes in Hubei Province, China, was $4.67 \times 10^9 m^3$ in 2019, equivalent to environmental benefits of $2.33 \times 10^6$ tons of carbon emission reduction, and economic benefits of $2.33 \times 10^7$ dollars.

Figure 1.1 shows that the predominant feedstock source of biogas or biomethane generation worldwide mainly consists of crop residues and animal manure. The Asia Pacific region has the largest production potential equivalent to over 200 million tons of oil equivalent (Mtoe) of biogas or biomethane by feedstock source. In the next two decades, it is projected that the overall potential of biogas production will grow rapidly because of the availability of substrate and the enhancement in waste collection and management, especially in developing economies (IEA, 2020).

### 1.3.3 Anaerobic Digestion Process

Generally, the anaerobic digestion process involves four stages (Figure 1.2). Meegoda et al. (2018) describe the different stages of the anaerobic digestion process. Each stage consists of four different types of microorganisms: hydrolysis (hydrolytic bacteria), acidogenesis (acidogens), acetogenesis (acetogens), and methanogenesis (methanogens). Hydrolysis is the initial step that converts insoluble

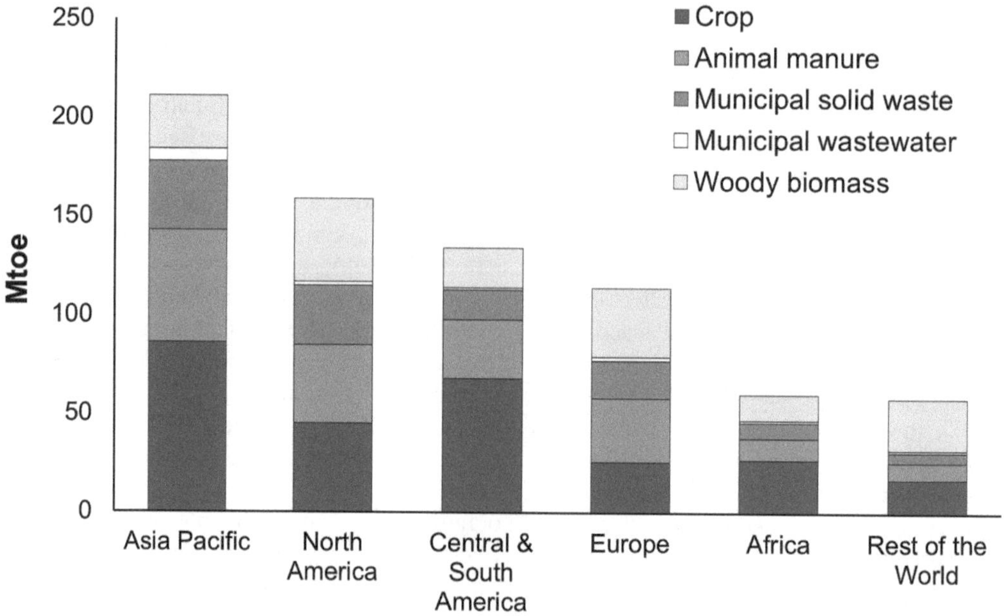


**FIGURE 1.1** Biogas or biomethane production by feedstock source in 2018. *Note*: C&S—Central and South America. Woody biomass feedstock—only for biomethane production. (Adapted from IEA, 2020.)

organics (macromolecules) into soluble organics (simpler molecules) by depolymerization or solubilization by the action of microbes. Chemically, hydrolysis means the breakdown of long-chain biomolecules in the presence of water. Biologically, hydrolysis is carried out by three main hydrolytic microorganisms: the proteolytic produces an enzyme known as a protease for the breakdown of proteins and peptides into ammonia and amino acids; the lipolytic generates lipases enzyme for the breakdown of saponifiable lipids into fatty acids and glycerol; and cellulolytic bacteria create hydrolases enzymes for the breakdown of polysaccharides into sugars. Most of these microorganisms are mostly obligate anaerobes and few of them are facultative. For solid substrates, hydrolysis is often the slowest among the four degradation stages, and hence pretreatment methods are necessary. Since the majority of hydrolysis products still require further degradation, the next step is acidogenesis.

In acidogenesis, soluble organic components including the products of hydrolysis are decomposed into volatile (organic) acids (propionic, butyric, valeric acid, etc.), ethanol, acetate, formate, carbon dioxide, and hydrogen by the action of acid-forming (fermentative) bacteria or acidogens. The products of acidogenesis are still large or inaccessible by methanogens, which are not ideal for methane production. Hence, they are subjected to acetogenesis. The products of acidogenesis (volatile organics, alcohols, etc.) are anaerobically digested during acetogenesis to produce acetic acid, hydrogen, and carbon dioxide. This conversion proceeds with the action of obligate hydrogen-producing acetogenic bacteria, which are considered acetogens and some are part of acidogens under syntrophy phenomena.

Methanogenesis, otherwise known as methane or methanogenic phase, is the last stage of anaerobic digestion. All of the intermediate products from the previous stages are converted into methane by the action of methanogens (*methanobacterium*, *methanobacillus*, *methanococcus*, and *methanosarcina*), which are strict anaerobes. Methanogens can be divided into two groups: acetate consumers that undertake aceticlastic methanogenesis, and hydrogen and carbon dioxide consumers that undertake hydrogenotrophic methanogenesis. The methanogenic substrates include acetate, hydrogen or carbon dioxide, formate, methanol, carbon monoxide, methylamines, methyl mercaptans,

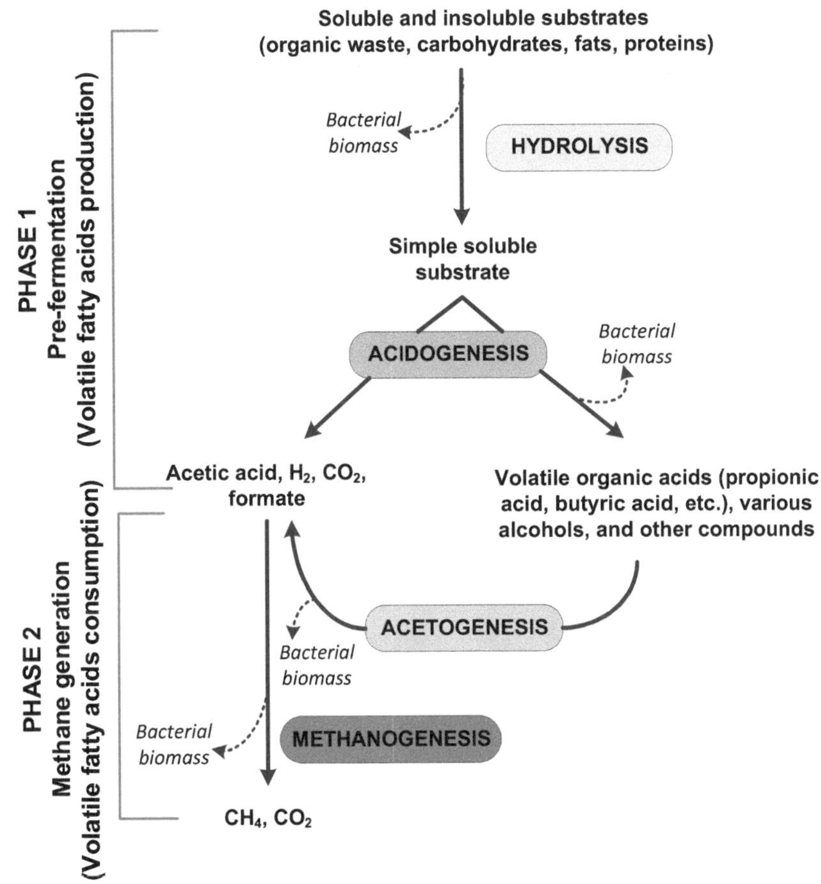


**FIGURE 1.2** Anaerobic digestion process. (Adapted with revision from Mata-Alvarez, 2003.)

and reduced metals that are converted to methane. Methane is produced either by means of the cleavage of acetic acid molecules to generate carbon dioxide and methane or by the reduction of carbon dioxide with hydrogen to yield methane and water. Methane-producing archaea generate large quantities of methane and carbon dioxide from acetic acid and also combine all of the available hydrogen gas with carbon dioxide to produce more methane. Methane production is higher from a reduction of carbon dioxide, but hydrogen concentration is limited in the digesters. This, therefore, makes acetate reaction to be the primary producer of methane. Sulfate-reducing bacteria are also present. They reduce sulfates and other sulfur compounds to hydrogen sulfide, most of which reacts with iron and other heavy metal salts to form insoluble sulfides, but there will always be some hydrogen sulfide remaining in the biogas. However, the ultimate yield of biogas depends on the composition and biodegradability of the waste feedstock, but its rate of production will depend on the population of bacteria and archaea, their growth conditions, and the temperature of the system.

### 1.3.4 Future Trends in Anaerobic Digestion

Advancements in anaerobic digestion technology in the 21st century deal with a growing amount of organic waste, the increasingly complex nature of substrate produced by our modern society, and the imperative sustainability and reduction in carbon emissions. Biogas and biomethane are renewable forms of energy that can provide a sustainable solution for the global energy crisis. The rising trend of the worldwide biogas production capacity indicates the high utilization of anaerobic digestion technology in treating organic waste (Figure 1.3). Furthermore, countries in Europe and Asia dominate biogas production (Figure 1.4).

Biogas produced from the anaerobic digestion of organic waste is mainly composed of methane and carbon dioxide with trace amounts of impurities. It offers a sustainable way to meet basic local energy needs for heating or cooking where access to national grids is limited. However, when biogas purity is elevated to biomethane by removing carbon dioxide and trace gases and densifying the methane concentration to enhance its calorific value, it can be used as a fuel for transportation, electricity generation, and a potential substitute for natural gas. Sahota et al. (2018) discussed several emerging technologies in biogas upgrading including water scrubbing, physical absorption, chemical absorption, pressure swing adsorption, membrane, cryogenic separation, in situ methane

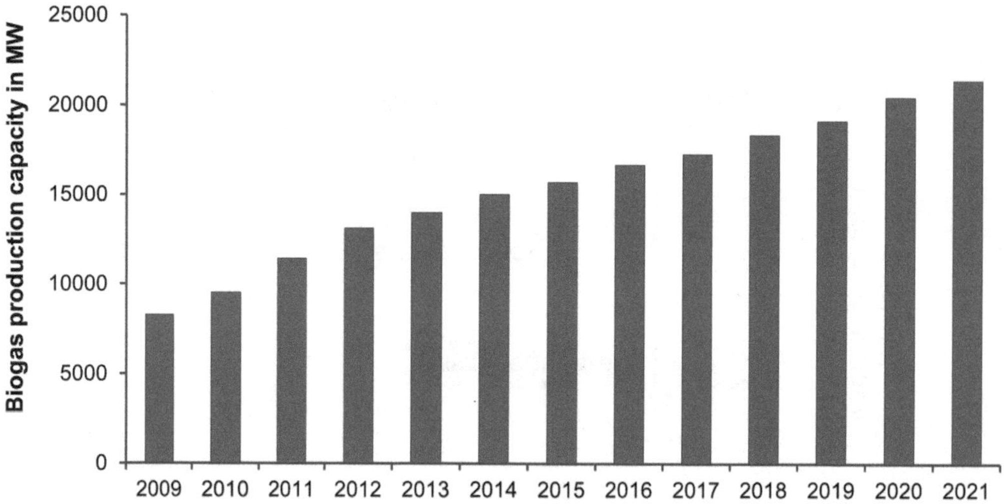


**FIGURE 1.3** Global biogas production capacity in MW (megawatts). (Adapted from Statista, 2022a.)

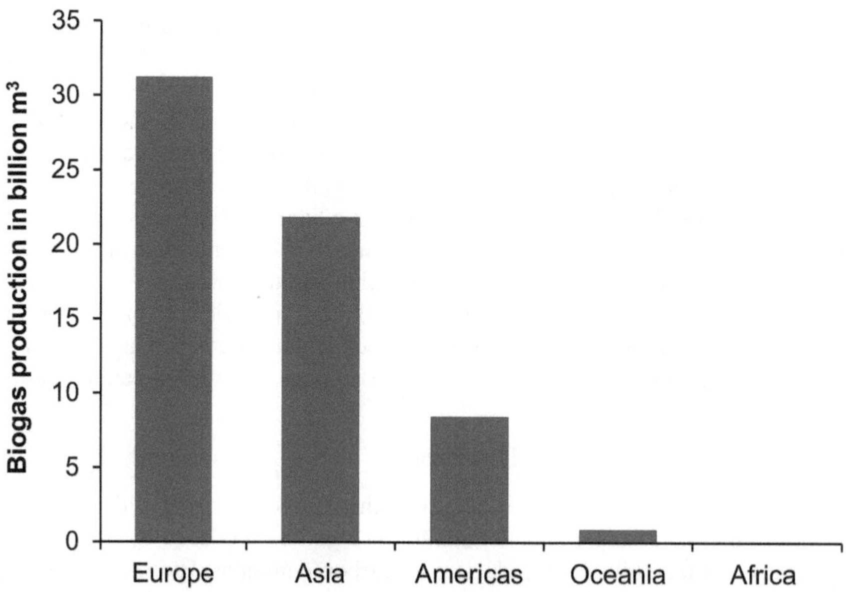


**FIGURE 1.4** Global biogas production in billion $m^3$. (Adapted from Statista, 2022b.)

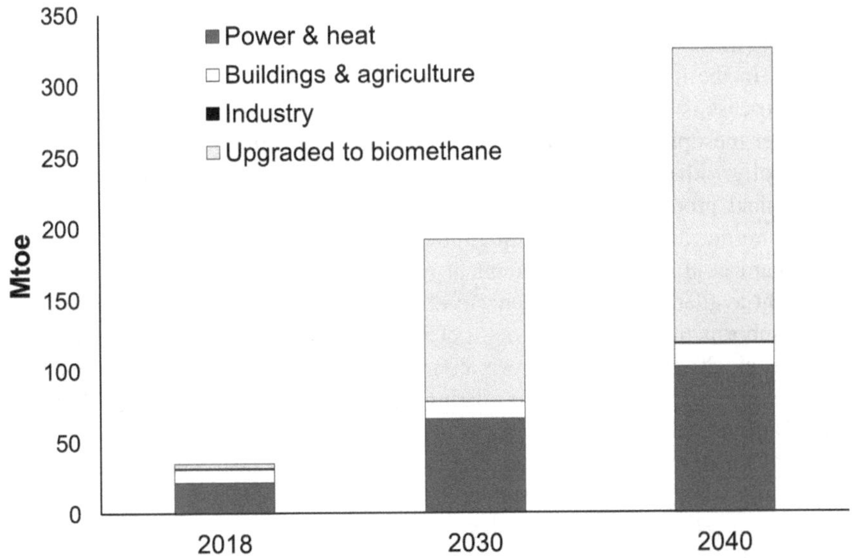


**FIGURE 1.5** Global biogas consumption by sector in the sustainable development scenario. (Adapted from IEA, 2020.)

enrichment, and hybrid technologies. IEA (2020) projected global biogas consumption, where a significant fraction of biogas production will be upgraded to biomethane (Figure 1.5).

Biogas upgrading through carbon dioxide utilization rather than its removal during the biogas upgrading process offers a suitable technology; carbon dioxide in biogas can be utilized as a feedstock for the synthesis of chemicals and fuels. The biogas upgrading technologies via carbon dioxide utilization include chemical and biological processes. Formerly, carbon dioxide was used to produce urea, methanol, cyclic carbonate, and salicylic acid. Thereafter, it was used to make carbon dioxide–based polymers and fuels. In the future, carbon dioxide is predicted to be used for producing carboxylic and succinic acid for the synthesis of chemicals, polymers, and solvents (Adnan et al., 2019).

There is a growing interest in the optimization of biohydrogen production via anaerobic digestion. Anaerobic digestion is a sustainable technology for organic waste management producing green energy (biogas, biomethane, and biohydrogen), stabilized waste residue used as bio-fertilizer, and mitigating greenhouse emissions. Hydrogen is a zero-carbon alternative energy source with a high calorific value (142 kJ/g) (Park et al., 2020; Trchounian, 2015). It can be obtained by steam reforming of methane gas, biomass gasification, and electrolysis of water and solar fuel cells (Khawer et al., 2022). Hydrogen is an intermediate product of the anaerobic digestion process that could be adjusted to enhance its production. Hydrogen production via anaerobic digestion specifically during the first two stages of hydrolysis and acidogenesis is also considered dark fermentation. Currently, there is a growing interest in enhancing biohydrogen production. Jokhio et al. (2022) demonstrated that co-digestion of banana plant waste (60%) and sewage sludge (40%) treated with 0.75 mol NaOH produced the highest yields of biomethane and biohydrogen. Furthermore, Sillero et al. (2022) revealed that the addition of 10 g/L of poultry manure to an equal proportion of sewage sludge and wine vinasse mixture improves the hydrogen production potential of 27.1 mL $H_2$/gVS added.

## 1.4 VERMICOMPOSTING

Vermicomposting is a low-cost sustainable waste disposal approach that has the potential of achieving a circular economy within waste management (Kamar Zaman & Yaacob, 2022; Alshehrei & Ameen, 2021; Yuvaraj et al., 2021). It is an environmentally sustainable biotechnology that can produce a peat-like, nutrient-absorbable product known as vermicompost for soil amelioration and seed germination (Mago et al., 2021; Usmani et al., 2019). The conventional composting process generally occurs in the thermophilic-controlled condition, whereas the bioconversion in the vermicomposting process is executed completely through the joint activities of microorganisms and earthworms under mesophilic conditions. The whole transformation process of organic residues is carried out naturally with the involvement of earthworms, making it a cost-efficient and energy-saving biotechnological process (Ghahdarijani et al., 2022). Compared to traditional compost, vermicompost is more nutritious with high quantities of N, P, and K (Boruszko, 2020), homogenous yield (Ghahdarijani et al., 2022) with excellent moisture-capturing ability (Cao et al., 2021), and long-term nutrient availability. In the scenario of farming proliferation and the continuous depletion of soil phosphorus due to soil erosion, a chemical fertilizer shortage is expected in the near future (Alewell et al., 2020). Therefore, vermicompost is a suitable organic alternative to traditional fertilizer, as it is not only capable of stimulating plant growth in agricultural (Kamar Zaman & Yaacob, 2022; Blouin et al., 2019) and horticultural practice (Álvarez et al., 2018) but also reducing the agrochemical fertilizer production. As the whole microbial decomposition process is totally chemical-free, the goals of sustainable development are secured. Hence, it is highly recommended to utilize vermicomposting, especially in developing countries, since it incurs less environmental impact (Shi et al., 2020), requires uncomplicated operation, and provides a useful eco-friendly quality post-product (i.e, vermicompost).

Vermicomposting is the decomposition process of organic waste through the joint activities of earthworms and microorganisms. Through the continuous activities of soil burrowing, feeding, and casting, the properties of organic litters are changed, releasing nutrients and minerals in the form available to plants. The decomposition process in vermicomposting is separated into two main phases known as the active phase and the maturation phase (Xia et al., 2019). In the beginning stage of vermicomposting, in the gut-associated process, earthworms play the key role in the conditioning and fragmentation of organic substrate. The earthworm gut, which contains numerous degradative enzymes, such as phosphatase, lipase, cellulase, and mannanase (Gusain & Suthar, 2020), will enhance the mineralization of organic fractions with the help of microflora. The easily ingested molecules in the organic substrate are utilized in this phase as the nourishment for the microbial community, whereas complex molecules will be degraded into simpler forms and blended into the substrates. Through the active phase, the intensive impact of earthworms will not only modify the physicochemical properties of organic matter but also increase the area for the habitation of microorganisms in the latter phase (Domínguez et al., 2019). In the second phase, the ingested organic compounds by earthworms will be biochemically decomposed until maturation by functional microorganisms (Xia et al., 2019). With the intensive activities of earthworms and microorganisms and large surface exposure area, the biochemical decomposition is accelerated, while essential nutrients such as nitrogen and phosphorus are retrieved and increased.

Vermicompost generated from the vermicomposting process has numerous benefits in retaining soil quality and promoting the growth and yield of plants via its enriched nutrient content (Blouin et al., 2019). Throughout the decomposition of organic matter by earthworms and microorganisms, the availability of micro- and macronutrients is increased, whereas plant hormones and humic acid are produced in matured vermicompost. Humic acid, which is the result of organic decay, is known to be an essential soluble component in modifying soil acidity for microbial activities and enhancing the water-holding capacity of the soil (Gholami et al., 2019). Humic acid acts as a plant-growth-supporting agent as it increases nutrient uptake through the chelate formation of the plant-essential elements such as Fe, Zn, Cu, Ca, Mg, N, P, and K, thereby increasing the enzymatic and microbial activities. Maji et al. have compared the ability of the humic acid–rich vermicompost and the conventional chemical fertilizer in supporting plant growth and photosynthetic pigments (Maji et al., 2017). The use of chemical fertilizer resulted in drastic microbial population loss, whereas the use of vermicompost convalesced soil improved microbial community structure and root nodulation (Castillo et al., 2016). Foliar application of humic acid–rich vermicompost in cabbage and tomato cultivation has proved to effectively boost the crop quality, growth, and yields (Goswami et al., 2017).

In addition to humic substances, vermicomposting also supplies macronutrients such as nitrogen and phosphorus to enhance plant growth. During vermicompost maturation, the availability of phosphorus and nitrogen increases significantly due to the involvement of phosphate-solubilizing bacteria and $N_2$ fixation bacteria (Busato et al., 2012). The rate of these ecological processes depends on the changes in microbial communities and types of earthworms in the soil. By increasing the number of rhizospheric bacteria, which are free-living microorganisms playing a predominant part in biological nitrogen fixation, vermicomposting accelerates the biological process and increases nutrient availability in soil. A study by Singh et al. has recorded the increment in $NH_4^+$-N and $NO_3^-$-N production from vermicomposting (Singh et al., 2012). The vermicomposting process also increases phosphate availability through the production of organic acid during decomposition (Busato et al., 2012). These acids aid in solubilizing phosphorous; hence, earthworms' involvement also accelerates the conversion of organic P into orthophates (Saha et al., 2008). Therefore, soil fertilized with vermicompost is nutrient-rich with high contents of ammonia, nitrates, and orthophosphates within. Chatterjee et al. have registered better soil aeration with enhanced bioavailability of nutrients, which boosted the yield of green bell peppers (Chatterjee et al., 2021).

Vermicomposting is a nonthermophilic process operating at temperatures lower than 35°C. This condition is most favorable for the intensive activities of earthworms; however, it is doubted that the ability to destruct pathogens will be affected. Notwithstanding, changes in the microbial community structure and population with interactions between earthworms and microbes have enhanced the pathogens' destruction capacity in vermicomposting (Shi et al., 2020). Another study has recorded the increment of salicylic acid in vermicompost (Domínguez et al., 2019). Salicylic acid, which is a plant hormone, helps regulate the growth of plants. Hence, it is a bio-regulator of plant–microbe interactions that induce biotic and abiotic stress tolerance (Maruri-López et al., 2019). Sahni et al. have shown that the synergetic effect of salicylic acid and vermicompost helps promote plant growth, thus reinforcing the barrier against chickpea wilt pathogens (Sahni et al., 2021). Similarly, various pathogens including *Salmonella* and *E. coli* were reported to rapidly reduce due to the activities of earthworms during vermicompost maturation (Procházková et al., 2018). Although not subjected in high-temperature conditions, the earthworm is the driver for the suppression of soil-borne diseases and pests during vermicomposting. Inside the earthworm gut, the oxygen level was reported to be low enough to damage the parasites in the waste (Monroy et al., 2009). After passing through the earthworm gut, the number of total coliforms in pig slurry was reduced significantly. With the presence of various intestinal enzymes inside the earthworm, bacteria cell walls are decomposed leading to pathogenic load reduction (Roubalová et al., 2020). Moreover, coelomic fluid in earthworms contains bioactive compounds such as antibacterial peptides and polysaccharides (Yakkou et al., 2021) that participate in the innate immunity against fungal diseases (Nadana et al., 2020).

Vermicomposting is an economical approach that has been used to convert organic residues and litter into nutrient-rich compost. During the feeding and burrowing process of earthworms, many physical changes occur, including the fragmentation and turnover of organic matter. When earthworms travel through the soil, they excrete the excreta of glutinous mucus and vermicast to the surrounding medium, which helps alter the properties of soil (Ahmad et al., 2021), enhancing bioremediation. Earthworm mucus, which is a water-soluble mixture of amino acids, carbohydrates, and glycoproteins, acts as an absorbent to induce humidification effects of organic matter (Bityutskii et al., 2012) and maintain the moisturizing level in the substrate. It also provides a protective barrier for the movement of earthworms in the soil. Mucus yield is high; therefore, the large amount of mucus generated can add to the nitrogen level in vermicompost. Earthworm intestine functions as a barrier against bacteria (Roubalová et al., 2020), thereby providing polysaccharide gums for increasing passage of water into the soil.

With the ability to generate microbially active secondary product via natural degradation, vermicomposting has been applied in the treatment of organic solid waste (Alshehrei & Ameen, 2021), sewage sludge (Ghahdarijani et al., 2022), dewatered sludge (Huang et al., 2022), and agricultural and industrial waste (Cao et al., 2021). It is considered as an ecologically sustainable waste management alternative to landfilling and has been utilized in New Zealand on a large scale for sewage sludge and paper mill sludge treatment to generate quality vermicompost for soil amendment (Quintern & Morley, 2017). Sewage sludge is known to be an abundant source of organic fertilizer due to its high nutrient content; however, it is also a microplastic (MP) container that can contaminate agricultural soil (Han et al., 2019). A study has concluded that the vermicomposting process can address the issue by significant reduction of MP content in sewage sludge (Ragoobur et al., 2022). Not only a natural MP controller, but vermicomposting also effectively remediates the noxious effects of heavy metals and destructs pathogens in the waste in order to generate high-quality vermicompost (Ameen & Al-Homaidan, 2022). According to Álvarez et al., the quality of the vermicompost was enhanced and the degradation process of organic matter was further accelerated when there was an additional amendment of biochar (Álvarez et al., 2018). Biochar is capable of boosting water-holding ability, increasing nutrient uptake of plants (Hijikata et al., 2015), meanwhile elevating pH conditions suitable for OM degradation by earthworms (Malińska et al., 2016). Recently, vermicomposting has become an attractive option for the treatment of food processing and agricultural waste (Ameen & Al-Homaidan, 2022; Mago et al., 2021; Gusain & Suthar, 2020). Leaf litter and crop residues such as leaves of bananas, cabbage, and cauliflower have the potential of being a resourceful fertilizer for plant growth and soil enrichment due to the richness in plant nutrients and biodegradable organic matters. However, the process of decomposing leaf waste is hindered because of the abundant presence of lignin, cellulose, and hemicellulose hinders the natural decay rate. Through vermicomposting, degradation rate is enhanced with the aid of lignocellulolytic in earthworm gut (Mago et al., 2021). Vermicomposting is also a sustainable option for treating industrial waste as it contains various plant nutrients. It was reported that distillery sludge waste with tea leaf residues could be used as a source of vermicompost production (Mahaly et al., 2018). On a lab scale, a combination of composting and vermicomposting has been utilized to stabilize poultry litter (Vicentin et al., 2021). Through both environmental and economic analysis, the strategy was proved to be beneficial and sustainable with the production of high-quality compost. It is clear that vermicompost has numerous benefits both to the environment and economy and can be applied to various waste types.

## 1.5 BIOREMEDIATION

Due to their hazardous impacts on both environment and humans, organic wastes attracted attention from all around the world. These pollutants were previously removed by expensive biochemical and physical methods that can create harmful byproducts (Ahmed et al., 2021). Therefore, progressive efforts have been made in recent literature to not only clean up the environment and reduce the risk to human health but also lessen the energy and expenses associated with waste treatment. Among them, sustainable remediation methods have become more significant in recent years with their main goal being to reduce the concentration of toxins. Besides, environmental protection has already been significantly influenced by the bio-based approach. As a result, bioremediation techniques have attracted a lot of attention and are considered as an economical and environmentally beneficial solution for the treatment of organic wastes (Cunningham et al., 2021). It is acknowledged as a green waste management method by the United States Environmental Protection Agency (US EPA).

Bioremediation is a process of using living systems or their enzymes to transform complicated pollutants into relatively safe or nontoxic chemical intermediates, and thus they can be used as nutrient-rich substrates for microbes as well as plants (Cunningham et al., 2021). When addressing the main organic waste streams, like sewage sludge, agricultural waste, or municipal waste (garden and food wastes), bioremediation strategies can be applied via composting or land farming combined with composting of biowaste. Microorganisms comprising some algae, fungi, or bacteria play a key role in bioremediation because they can degrade, detoxify, immobilize, or stabilize a variety of toxic and biodegradable contaminants from the environment, turning them into less dangerous forms (Patel et al., 2022). In fungal bioremediation of oil-contaminated soils, bacteria (57%), enzymes (19%), fungi (13%), algae (6%), plants (4%), and protozoa represent the majority of the bioremediation agents (Quintella et al., 2019). Thanks to microorganism activities, bioremediation is also capable of treating not only biodegradable contaminants but also recalcitrant organic pollutants with poor biodegradability in industrial, agricultural, and municipal waste fields, such as phenols, endocrine-disrupting chemicals, chlorinated phenols, petroleum hydrocarbons, azo dyes, pesticides, and other persistent organic pollutants. A study by Gómez et al. used *Abarenicola pusilla* to treat marine aquaculture sludge and showed a high removal rate of total organic matter of ~85% (Gómez et al., 2019). The degradation of p-nitrophenol by *Pseudomonas* sp. was reported in a previous research with an effectiveness of 90% for the p-nitrophenol range of 85 ppm (Das & Dey, 2020). The effectiveness of bioremediation (>60%) for soils contaminated by a variety of petroleum hydrocarbons has been shown in multiple research (Gielnik et al., 2019; Ahmadi et al., 2016; Naowasarn & Leungprasert, 2016). Through the primary ways of the terminal, subterminal, ω-and β-oxidations, *n*-alkanes are broken down and converted into a carbon source for the bacteria, leading to the reduction of petroleum hydrocarbon concentrations in contaminated soils (Varjani, 2017). In addition, as an eco-friendly approach for environmental sustainability, bioremediation has the capability to degrade residual pesticides using bacteria (e.g., *Pseudomonas* sp., *Bacillus* sp., and *Klebsiella* sp.), fungi (e.g., *Trichoderma* sp., *Aspergillus* sp., and white-rot fungi), and algae (e.g., *Chlamydomonas* and marine *Chlorella*) (Sehrawat et al., 2021). As an example of organophosphate insecticides, numerous bacterial strains have been identified and shown to be capable of metabolizing and/or co-metabolizing these pesticides in the environment (Mulla et al., 2020). Recent research has also applied various white-rot fungi to treat 2,3,7,8-tetrachlorodibenzo-p-dioxin and indicated up to 73% degradation rate after 28 days for a species of *Rigidoporus* (Dao et al., 2019). Since the enzymes produced by these ligninolytic fungi can degrade dioxin as well as polycyclic aromatic hydrocarbons, polar and water-soluble byproducts of the breakdown process can be then accessed by fungal and other microbial metabolisms in that environment (Lin et al., 2022). This helps to prevent dangerous compounds from entering the environment and harming both the ecology and human health.

To degrade or detoxify organic wastes, as described above, the bioremediation process employs a variety of microorganisms with varied metabolic pathways. In general, these microorganisms use inorganic and organic substances as a source of nutrients to thrive and degrade pollutants. These wastes are then transformed into inorganic substances during the bioremediation process, such as carbon dioxide, methane, and water. Depending on the location of the polluted site and the features, kind, and severity of the pollutants, the bioremediation process is carried out either in situ (bioremediation process at the site) or ex situ (bioremediation process elsewhere) (Figure 1.6). It is crucial to correctly implement the process under in situ or ex situ ways in order to achieve effective organic waste treatment as well as the greatest possible outcomes in terms of the economy and environment. A variety of remediation processes are traditionally included in bioremediation, including bioaugmentation, which is a natural attenuation process using native microorganisms to be able to mineralize pollutants and then use them as a nutrient source; biostimulation, which is the use of nutrients (e.g., nitrogen, phosphorus, or biosurfactants to microorganisms); phytoremediation, which uses some plants; and biomineralization, which is the thorough biodegradation of organic substances into inorganic components. Different bioremediation techniques are illustrated in Figure 1.6.

Although bioremediation is an environmentally friendly and cost-effective approach, microbial breakdown in the natural environment has little effect on the molecular structure of persistent pollutants, which results in their low clearance (Peng et al., 2020). Only low initial concentrations can be successfully treated since extremely hazardous substances like polychlorinated dibenzo-p-dioxins and furans take longer to decompose (Lin et al., 2022). Therefore, numerous innovative efforts have been undertaken to increase the effectiveness of removing these persistent contaminants from

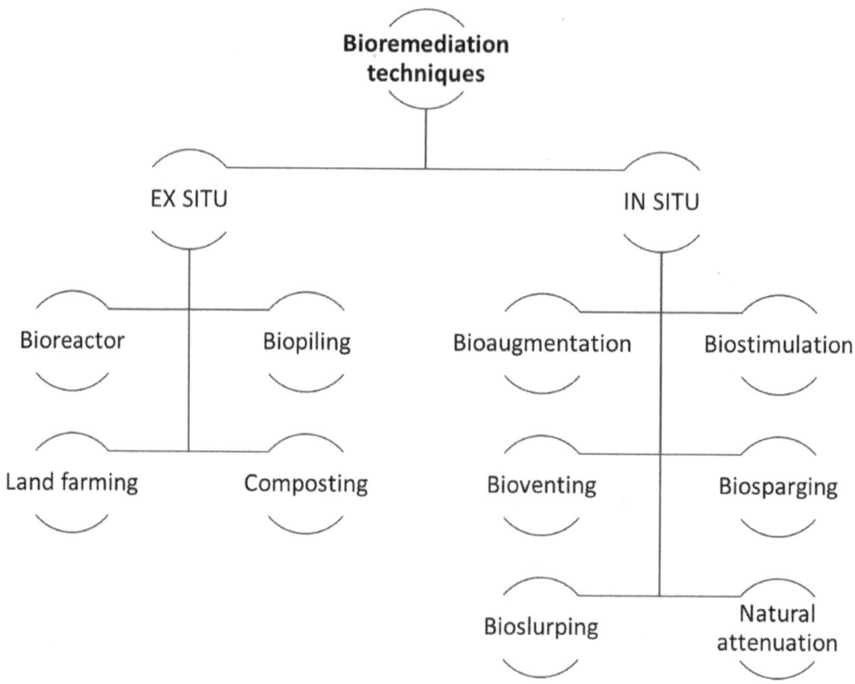


**FIGURE 1.6** Classification of various bioremediation techniques. (Adapted from Patel et al., 2022.)

the environment (Sebastian & Rao, 2022; Hung et al., 2017). Incorporating the latest emerging technologies, it is feasible to increase the biodegradation capacity of the used microorganisms, thereby increasing the efficacy of bioremediation. These technologies comprise electrokinetics, microbial fuel cells (MFCs), and nanotechnology. For electrokinetic-bioremediation technology, electrolysis, electrophoresis, electroosmosis, electromigration, and diffusion are some of the phenomena in electrokinetics to orient and accelerate the transport of pollutants and microbes for bioremediation (Saxena et al., 2021). Wu et al. investigated an electrokinetic-bioremediation system for n-hexadecane degradation with the presence of ryegrass (Wu et al., 2020). The examination of the microbial community revealed that gram-positive bacteria were largely responsible for the breakdown of n-hexadecane while a new microbial community was progressively formed in the rhizosphere that continued to metabolize n-hexadecane. The electro-bioremediation of phenanthrene – a polycyclic aromatic hydrocarbon – was reported in previous research using *Mycobacterium pallens* and showed its high removal from the soil (Hassan et al., 2017). MFC as a bio-electrochemical device allows microbial activity to generate electricity using organic substrate (Verma et al., 2021). Organic substances are oxidized by microorganisms immobilized at the MFC's anode in order to release protons and electrons that flow to MFC's cathode to produce current. The reduced compound formation is caused by the cathode's reaction of protons and electrons with electron acceptors. Christwardana et al. (2020) evaluated the performance of yeast-assisted microalgal MFCs for the bioremediation of cafeteria wastewater. The system showed a high removal rate of organic compounds with more than 60% COD (chemical oxygen demand) removal. Another research pointed out up to 90% degradation rate of sodium benzoate in a MFC system using a novel consortium using five electrogenic bacteria, namely, *Kocuria rosea* (GTPAS76), two strains of *Bacillus circulans* (GTPO28 and GTPAS54), and two strains of *Corynebacterium vitaeruminis* (GTPO38 and GTPO42) (Mukherjee et al., 2021). The development of nanotechnology and the use of nanoparticles in combination with bioremediation constitute another unique approach to pushing bioremediation beyond its limitations on low effectiveness (Hidangmayum et al., 2022; Pandey et al., 2022). Microbes can act as a catalyst since their enzyme is present for a specific reaction to inorganic nanoparticles. Bacterial nanoparticles produced can be used to detoxify and bioremediate soil, water, and other environments in highly polluted conditions (Ramezani et al., 2021). Parthipan et al. indicated that the combination between iron nanoparticles and biosurfactants produced by *Bacillus subtilis* A1 resulted in the high degradation rate of 85% of the mixture of PAHs including anthracene, pyrene, and benzo(a)pyrene (Parthipan et al., 2022).

## 1.6 PERSPECTIVES

The emerging green technologies for organic waste treatment discussed in this chapter include fermentation, anaerobic digestion, vermicomposting, and bioremediation. Fermentation utilizes microorganisms in converting carbonaceous organic materials into sustainable products like bioplastics. Bioplastic is a bio-based, biodegradable plastic, or both, and is considered one of the alternatives to the plastic problem. Bio-based plastics or partially bio-based can be nonbiodegradable plastics such as bio-based PP or PET or can be biodegradable such as PLA or PHA. PLA and PHA are produced via fermentation using the organic waste stream. However, more research still needs to be done prior to its commercialization.

Anaerobic digestion is a natural process of biomass degradation in the absence of oxygen. It is considered a sustainable technology for organic waste treatment producing green energy (biogas, biomethane, and hydrogen) and nutrient-rich digestate residue while reducing methane emissions. Advances in anaerobic digestion technology in the 21st century is crucial to deal with the ever-increasing amount and complexity of organic waste produced by our modern society, the imperative sustainability, and reduction in carbon emissions. Biogas, biomethane, and biohythane are renewable forms of energy that can provide a sustainable solution for the global energy crisis. Several emerging technologies in biogas upgrading include water scrubbing, physical absorption, chemical absorption, pressure swing adsorption, membrane, cryogenic separation, in situ methane enrichment, and hybrid technologies.

Vermicomposting is a biotechnological waste disposal approach to decompose organic compounds through the joint activities of earthworms and microorganisms. It is capable of generating a peat-like, nutrient-absorbable product that aids soil amelioration and seed germination. Compared to the traditional composting process, vermicomposting has the increased availability of macronutrients and plant hormones that are necessary for modifying soil quality for boosting plant growth. The whole process is operated in the nonthermophilic condition, which is both economically energy-saving and favorable for microbial activities; hence, with the changes in microbial community structure and population, the pathogen destruction ability is also secured. Since it is a sustainable eco-friendly approach, vermicomposting is employed in the treatment of not only organic solid waste but also in industrial waste and sewage sludge treatment.

Bioremediation can be employed as nutrient-rich substrates for both microorganisms and plants by utilizing living systems or their enzymes to transform complex contaminants into relatively safe or nontoxic chemical intermediates. It is possible to treat recalcitrant organic pollutants with poor biodegradability in industrial, agricultural, and municipal waste areas in addition to biodegradable contaminants thanks to microorganism activities. This aids in preventing hazardous substances from hurting the ecology and public health by entering the environment. Despite being considered as an economically and environmentally benign method, microbial breakdown in the natural environment has little impact on the molecular structure of persistent contaminants. By incorporating nanotechnology, electrokinetics, or MFCs, it is feasible to increase the biodegradation capacity of the used microorganisms, thus boosting the effectiveness of bioremediation.

## 1.7 SUMMARY AND CONCLUSIONS

The advances in industrialization and high resource consumption require a sustainable supply of resources and the utilization of the generated waste. Organic waste is a dominant type of waste generated in our society in spite of technological advances. The organic nature of waste is viewed as a potentially valuable resource once treated sustainably. However, organic waste treatment remains a global challenge in the 21st century. Emerging green technologies transforming organic waste into valuable resources are a sustainable approach to mitigating climate change and conserving our depleting resources. Technologies such as fermentation, anaerobic digestion, vermicomposting, and bioremediation are widely utilized in converting organic waste into sustainable resources. The fermentation process utilizes organic waste to produce valuable products such as bioplastics and is considered one of the solutions to the plastic problem. Anaerobic digestion technology is a sustainable process for organic waste treatment generating renewable form of energy. Advancements in anaerobic digestion focus on densifying methane content to enhance its calorific value. Hydrogen is an intermediate product of the anaerobic digestion process that could be adjusted to enhance its production. Vermicomposting is another sustainable eco-friendly approach to organic waste treatment utilizing earthworms and microorganisms to produce a peat-like bio-fertilizer or vermicompost. Vermicomposting is not limited to organic solid waste, but it is also employed for industrial waste and sewage sludge treatment. Bioremediation is considered an economically and environmentally benign method used to treat recalcitrant organic pollutants with poor biodegradability. It is feasible to enhance the biodegradation capability of the utilized microorganisms and improve the efficiency of bioremediation. These emerging green technologies for organic waste treatment and management support the circular economy to tackle global challenges of climate change, depleting resources, and environmental pollution.

## REFERENCES


Abbassi-Guendouz, A., Brockmann, D., Trably, E., Dumas, C., Delgenès, J-P. Steyer, J-P., Escudié, R. 2012. Total solids content drives high solid anaerobic digestion via mass transfer limitation. *Bioresource Technology*, 111, 55–61. doi:10.1016/j.biortech.2012.01.174.

Acosta-Cárdenas, Alejandro, Wilman Alcaraz-Zapata, Mariana Cardona-Betancur. 2018. Sugarcane molasses and vinasse as a substrate for polyhydroxyalkanoates (PHA) production. *Dyna*, 85(206), 220–225.

Adnan, A.I., Ong, M.Y., Nomanbhay, S., Chew, K.W., Show, P.L. 2019. Technologies for biogas upgrading to biomethane: A review. *Bioengineering*, 6(92). doi:10.3390/bioengineering6040092

Ahmad, A., Aslam, Z., Bellitürk, K., Iqbal, N., Naeem, S., Idrees, M., Kaleem, Z., Nawaz, M.Y., Nawaz, M., Sajjad, M. 2021. Vermicomposting methods from different wastes: An environment friendly, economically viable and socially acceptable approach for crop nutrition: A review. *International of Agriculture and Food Science*, 5(1), 58–68.

Ahmadi, M., Dashtestani, M., Takdastan, A., Jorfi, S., Ramavandi, B. 2016. Data on biodegradation of total petroleum hydrocarbons using co-composting of cow manure/oily drill wastes. *Data in Brief*, 9, 723–726.

Ahmed, S.F., Mofijur, M., Nuzhat, S., Chowdhury, A.T., Rafa, N., Uddin, M.A., Inayat, A., Mahlia, T.M.I., Ong, H.C., Chia, W.Y., Show, P.L. 2021. Recent developments in physical, biological, chemical, and hybrid treatment techniques for removing emerging contaminants from wastewater. *Journal of Hazardous Materials*, 416, 125912.

Alewell, C., Ringeval, B., Ballabio, C., Robinson, D.A., Panagos, P., Borrelli, P. 2020. Global phosphorus shortage will be aggravated by soil erosion. *Nature Communications*, 11(1), 1–12.

Alshehrei, F., Ameen, F. 2021. Vermicomposting: A management tool to mitigate solid waste. *Saudi Journal of Biological Sciences*, 28(6), 3284–3293.

Álvarez, J.M., Pasian, C., Lal, R., López, R., Díaz, M.J., Fernández, M. 2018. Morpho-physiological plant quality when biochar and vermicompost are used as growing media replacement in urban horticulture. *Urban Forestry & Urban Greening*, 34, 175–180.

Ameen, F., Al-Homaidan, A.A. 2022. Improving the efficiency of vermicomposting of polluted organic food wastes by adding biochar and mangrove fungi. *Chemosphere*, 286, 131945.

Bhatia, S.K., Otari, S.V., Jeon, J.-M., Gurav, R., Choi, Y.-K., Bhatia, R.K., Pugazhendhi, A., Kumar, V., Banu, J.R., Yoon, J.-J. 2021. Biowaste-to-bioplastic (polyhydroxyalkanoates): Conversion technologies, strategies, challenges, and perspective. *Bioresource Technology*, 326, 124733.

Bishop, G., Styles, D., Lens, P.N.L. (2021). Environmental performance of bioplastic packaging on fresh food produce: A consequential life cycle assessment, *Journal of Cleaner Production*, 317, 128377.

Bityutskii, N., Maiorov, E., Orlova, N. 2012. The priming effects induced by earthworm mucus on mineralization and humification of plant residues. *European Journal of Soil Biology*, 50, 1–6.

Blouin, M., Barrere, J., Meyer, N., Lartigue, S., Barot, S., Mathieu, J. 2019. *Vermicompost Significantly Affects Plant Growth: A Meta-Analysis*, Vol. 39, Springer, New York, pp. 1–15.

Boethius, A. 2016. Something rotten in Scandinavia: The world's earliest evidence of fermentation. *Journal of Archaeological Science*, 66, 169–180.

Bolzonella, D., Battista, F., Cavinato, C., Gottardo, M., Micolucci, F., Pavan, P. 2019. Biohythane production from food wastes. In: A. Pandey, S. Venkata Mohan, J.-S. Chang, P.C. Hallenbeck, C. Larroche (eds), *Biomass, Biofuels, Biochemicals, Biohydrogen*, 2nd edn. Elsevier, Amsterdam, pp. 347–368. doi:10.1016/B978-0-444-64203-5.00013-7.

Boruszko, D. 2020. Vermicomposting as an alternative method of sludge treatment. *Journal of Ecological Engineering*, 21, 2.

BTA-Biotechnical Recycling. 2022. Anaerobic wet digestion of the organic fraction of waste. https://bta-international.de/en/technologien/anaerobe-nassvergaerung-der-organikfraktion-von-abfaellen

Burniol-Figols, A., Varrone, C., Le, S.B., Daugaard, A.E., Skiadas, I.V., Gavala, H.N. 2018. Combined polyhydroxyalkanoates (PHA) and 1, 3-propanediol production from crude glycerol: Selective conversion of volatile fatty acids into PHA by mixed microbial consortia. *Water Research*, 136, 180–191.

Busato, J.G., Lima, L.S., Aguiar, N.O., Canellas, L.P., Olivares, F.L. 2012. Changes in labile phosphorus forms during maturation of vermicompost enriched with phosphorus-solubilizing and diazotrophic bacteria. *Bioresource Technology*, 110, 390–395.

Calbry-Muzyka, A., Madi, H., Rüsch-Pfund, F., Gandiglio, M., Biollaz, S. 2022. Biogas composition from agricultural sources and organic fraction of municipal solid waste. *Renewable Energy*, 181, 1000–1007.

Cao, Y., Tian, Y., Wu, Q., Li, J., Zhu, H. 2021. Vermicomposting of livestock manure as affected by carbon-rich additives (straw, biochar and nanocarbon): A comprehensive evaluation of earthworm performance, microbial activities, metabolic functions and vermicompost quality. *Bioresource Technology*, 320, 124404.

Castillo, J.M., Beguet, J., Martin-Laurent, F., Romero, E. 2016. Multidisciplinary assessment of pesticide mitigation in soil amended with vermicomposted agroindustrial wastes. *Journal of Hazardous Materials*, 304, 379–387.

Chatterjee, D., Dutta, S.K., Kikon, Z.J., Kuotsu, R., Sarkar, D., Satapathy, B., Deka, B.C. 2021. Recycling of agricultural wastes to vermicomposts: Characterization and application for clean and quality production of green bell pepper (*Capsicum annuum* L.). *Journal of Cleaner Production*, 315, 128115.

Christwardana, M., Hadiyanto, H., Motto, S.A., Sudarno, S., Haryani, K. 2020. Performance evaluation of yeast-assisted microalgal microbial fuel cells on bioremediation of cafeteria wastewater for electricity generation and microalgae biomass production. *Biomass and Bioenergy*, 139, 105617.

Corless, V. 2022. How biogas from human waste will lead to energy independence. https://www.advancedsciencenews.com/how-biogas-from-human-waste-will-lead-to-energy-independence/

Crespy, D., Bozonnet, M., Meier, M. 2008. 100 years of bakelite, the material of a 1000 uses. *Angewandte Chemie International Edition*, 47(18), 3322–3328.

Cunningham, C., Peshkur, T., Kuyukina, M., Ivshina, I. 2021. Sustainable bioremediation of hydrocarbon contaminated soils: Opportunities for symbiosis with organic waste management? *Russian Journal of Ecology*, 52(6), 463–469.

Dao, A.T.N., Vonck, J., Janssens, T.K.S., Dang, H.T.C., Brouwer, A., de Boer, T.E. 2019. Screening white-rot fungi for bioremediation potential of 2,3,7,8-tetrachlorodibenzo-p-dioxin. *Industrial Crops and Products*, 128, 153–161.

Das, A., Dey, A. 2020. P-nitrophenol-bioremediation using potent pseudomonas strain from the textile dye industry effluent. *Journal of Environmental Chemical Engineering*, 8(4), 103830.

Department for Environment, Food and Rural Affairs (DEFRA). 2011. *Anaerobic Digestion Strategy and Action Plan*. Nobel House, 17 Smith Square, London. https://assets.publishing.service.gov.uk/government/uploads/system/uploads/attachment_data/file/69400/anaerobic-digestion-strat-action-plan.pdf

Deublein, D., Steinhauser, A. 2011. *Biogas from Waste and Renewable Resources: An Introduction*. Wiley – VCH, New York.

Domínguez, J., Aira, M., Kolbe, A.R., Gómez-Brandón, M., Pérez-Losada, M. 2019. Changes in the composition and function of bacterial communities during vermicomposting may explain beneficial properties of vermicompost. *Scientific Reports*, 9(1), 1–11.

Emetere, M.E., Chikwendu, L., Afolalu, S.A. 2022. Improved biogas production from human excreta using chicken feather powder: A sustainable option to eradicating poverty. *Global Challenges*, 6, 2100117. https://onlinelibrary.wiley.com/doi/epdf/10.1002/gch2.202100117.

European Bioplastics. 2021. Bioplastic materials. https://www.european-bioplastics.org/bioplastics/materials/, accessed January 12, 2023.

European Bioplastics. 2022. Bioplastics market data. https://www.european-bioplastics.org/market/#, accessed January 12, 2023.

Fauzi, A.H.M., Chua, A.S.M., Yoon, L.W., Nittami, T., Yeoh, H.K. 2019. Enrichment of PHA-accumulators for sustainable PHA production from crude glycerol. *Process Safety and Environmental Protection*, 122, 200–208.

FoodPrint. 2019. Can food waste become biodegradable plastics? https://foodprint.org/blog/biodegradable-plastic-food-waste/

Ghahdarijani, A.R.J., Hoodaji, M., Tahmourespour, A. 2022. Vermicomposting of sewage sludge with organic bulking materials to improve its properties. *Environmental Monitoring and Assessment*, 194(8), 1–14.

Gholami, H., Ghani, A., Fard, F.R., Saharkhiz, M.J., Hazrati, H. 2019. Changes in photosynthetic pigments and uptake of some soil elements by chicory supplied with organic fertilizers. *Acta Ecologica Sinica*, 39(3), 250–256.

Gielnik, A., Pechaud, Y., Huguenot, D., Cébron, A., Esposito, G., van Hullebusch, E.D. 2019. Bacterial seeding potential of digestate in bioremediation of diesel contaminated soil. *International Biodeterioration & Biodegradation*, 143, 104715.

Gómez, S., Hurtado, C.F., Orellana, J. 2019. Bioremediation of organic sludge from a marine recirculating aquaculture system using the polychaete *Abarenicola pusilla* (Quatrefages, 1866). *Aquaculture*, 507, 377–384.

Goswami, L., Nath, A., Sutradhar, S., Bhattacharya, S.S., Kalamdhad, A., Vellingiri, K., Kim, K.-H. 2017. Application of drum compost and vermicompost to improve soil health, growth, and yield parameters for tomato and cabbage plants. *Journal of Environmental Management*, 200, 243–252.

Gottardo, M., Bolzonella, D., Tuci, G.A., Valentino, F., Majone, M., Pavan, P., Battista, F. 2022. Producing volatile fatty acids and polyhydroxyalkanoates from foods by-products and waste: A review. *Bioresource Technology*, 361, 127716.

Gusain, R., Suthar, S. 2020. Vermicomposting of duckweed (Spirodela polyrhiza) by employing Eisenia fetida: Changes in nutrient contents, microbial enzyme activities and earthworm biodynamics. *Bioresource Technology*, 311, 123585.

Han, X., Lu, X., Vogt, R.D. 2019. An optimized density-based approach for extracting microplastics from soil and sediment samples. *Environmental Pollution*, 254, 113009.

Hassan, I., Mohamedelhassan, E., Yanful, E.K., Yuan, Z.-C. 2017. Solar power enhancement of electrokinetic bioremediation of phenanthrene by *Mycobacterium pallens*. *Bioremediation Journal*, 21(2), 53–70.

Hidangmayum, A., Debnath, A., Guru, A., Singh, B., Upadhyay, S., Dwivedi, P. 2022. Mechanistic and recent updates in nano-bioremediation for developing green technology to alleviate agricultural contaminants. *International Journal of Environmental Science and Technology*, 20, 11693–11718.

Hijikata, N., Yamauchi, N., Ishiguro, M., Ushijima, K., Funamizu, N. 2015. Suitability of biochar as a matrix for improving the performance of composting toilets. *Waste Management & Research*, 33(4), 313–321.

Huang, K., Zhang, Y., Xu, J., Guan, M., Xia, H. 2022. Feasibility of vermicomposting combined with room drying for enhancing the stabilization efficiency of dewatered sludge. *Waste Management*, 143, 116–124.

Hung, W., Huang, W.-Y., Lin, C., Vu, C.T., Yotapukdee, S., Kaewlaoyoong, A., Chen, J.-R., Shen, Y.-H. 2017. The use of ultrasound-assisted anaerobic compost tea washing to remove poly-chlorinated dibenzo-p-dioxins (PCDDs), dibenzo-furans (PCDFs) from highly contaminated field soils. *Environmental Science and Pollution Research*, 24(23), 18936–18945.

International Energy Agency (IEA). 2020. *Outlook for Biogas and Biomethane: Prospects for Organic Growth*. IEA, Paris. https://www.iea.org/reports/outlook-for-biogas-and-biomethane-prospects-for-organic-growth

Jain, S., Jain, S., Tim, I., Lee, J., Wah, Y. 2015. A comprehensive review on operating parameters and different pretreatment methodologies for anaerobic digestion of municipal solid waste. *Renewable and Sustainable Energy Reviews*, 52, 142–154, 10.1016/j.rser.2015.07.091

Jokhio, K., Qureshi, S.S., Sahito, A.R., Qureshi, M., Lashari, I.A., Mubarak, N.M., Tripathi, M., Al-Mohaimeed, A.M., Elshikh, M.S. 2022. Enhanced bio-methane and bio-hydrogen production using banana plant waste and sewage sludge through anaerobic co-digestion, *International Journal of Hydrogen Energy*. doi:10.1016/j.ijhydene.2022.04.238.

Juanga-Labayen, J., Yanac, K., Yuan, Q. 2021. Effect of substrate-to-inoculum ratio on anaerobic digestion of treated and untreated cotton textile waste. *International Journal of Environmental Science and Technology*, 18, 287–296. doi:10.1007/s13762-020-02831-9

Juanga-Labayen, J.P., Labayen, I.V., Yuan, Q. 2022. A review on textile recycling practices and challenges. *Textiles*, *2*(1), 174–188. doi:10.3390/textiles2010010

Kamar Zaman, A.M., Yaacob, J.S. 2022. Exploring the potential of vermicompost as a sustainable strategy in circular economy: Improving plants' bioactive properties and boosting agricultural yield and quality. *Environmental Science and Pollution Research*, 1–17. doi:10.1007/s11356-021-18006-z.

Kampen, W.H. 2014. Nutritional requirements in fermentation processes. In: H.C. Vogel, C.M. Todaro (eds), *Fermentation and Biochemical Engineering Handbook*. Elsevier, Amsterdam, pp. 37–57.

Kaza, S., Bhada-Tata, P. 2018. Decision Maker's Guides for Solid Waste Management Technologies. Urban Development Series Knowledge Papers. World Bank, Washington, DC. https://openknowledge.worldbank.org/handle/10986/31694

Kaza, S., Yao, L.C.; Bhada-Tata, P., Van Woerden, F. 2018. What a Waste 2.0: A Global Snapshot of Solid Waste Management to 2050. Urban Development, World Bank, Washington, DC. https://openknowledge.worldbank.org/handle/10986/30317

Khalid A., Arshad M., Anjum M., Mahmood T., Dawson L. 2011. The anaerobic digestion of solid organic waste-Review. *Waste Management*, 31(8), 1737–1744

Khawer, M.U.B., Naqvi, S.R., Ali, I., Arshad, M., Juchelková, D., Anjum, M.W., Naqvi, M. 2022. Anaerobic digestion of sewage sludge for biogas and biohydrogen production: State-of-the-art trends and prospects. *Fuel*, 329, 125416. doi:10.1016/j.fuel.2022.125416.

Komilis, D., Barrena, R., Grando, R.L., Vogiatzi, V., Sánchez, A., Font, X. 2017. A state of the art literature review on anaerobic digestion of food waste: Influential operating parameters on methane yield. *Reviews in Environmental Science and Biotechnology*, 16(2), 1–14.

Kosseva, M.R. Edy, R. 2018. Trends in the biomanufacture of polyhydroxyalkanoates with focus on downstream processing. *International Journal of Biological Macromolecules*, 107, 762–778.

Kumar, A., Samadder, S.R. 2020. Performance evaluation of anaerobic digestion technology for energy recovery from organic fraction of municipal solid waste: A review, *Energy*, 197, 117253. doi:10.1016/j.energy.2020.117253.

Li, Y., Park, S.Y., Zhu, J. 2011. Solid-state anaerobic digestion for methane production from organic waste. *Renewable and Sustainable Energy Reviews*, 15(1), 821–826. doi:10.1016/j.rser.2010.07.042.

Lin, C., Cheruiyot, N.K., Bui, X.-T., Ngo, H.H. 2022. Composting and its application in bioremediation of organic contaminants. *Bioengineered*, 13(1), 1073–1089.

Liu, L., Wang, J., Rosenberg, D., Zhao, H., Lengyel, G., Nadel, D. 2018. Fermented beverage and food storage in 13,000 y-old stone mortars at Raqefet Cave, Israel: Investigating Natufian ritual feasting. *Journal of Archaeological Science: Reports*, 21, 783–793.

Liu, T., Ferrari, G., Pezzuolo, A., Alengebawy, A., Jin, K., Yang, G., Li, Q., Ai, P. 2023. Evaluation and analysis of biogas potential from agricultural waste in Hubei Province, China. *Agricultural Systems*, 205, 103577. doi:10.1016/j.agsy.2022.103577.

Logan, M., Visvanathan, C. 2019. Management strategies for anaerobic digestate of organic fraction of municipal solid waste: Current status and future prospects. *Waste Management & Research*, 37, 27–39. doi:10.1177/0734242X18816793

Lonnqvist, T., Sanches-Pereira, A., Sandberg, T. 2015. Biogas potential for sustainable transport: A Swedish regional case. *The Journal of Cleaner Production*, 108, 1105–1114. doi:10.1016/j.jclepro.2015.07.036.

Mago, M., Yadav, A., Gupta, R., Garg, V. 2021. Management of banana crop waste biomass using vermicomposting technology. *Bioresource Technology*, 326, 124742.

Mahaly, M., Senthilkumar, A.K., Arumugam, S., Kaliyaperumal, C., Karupannan, N. 2018. Vermicomposting of distillery sludge waste with tea leaf residues. *Sustainable Environment Research*, 28(5), 223–227.

Maji, D., Misra, P., Singh, S., Kalra, A. 2017. Humic acid rich vermicompost promotes plant growth by improving microbial community structure of soil as well as root nodulation and mycorrhizal colonization in the roots of Pisum sativum. *Applied Soil Ecology*, 110, 97–108.

Malińska, K., Zabochnicka-Świątek, M., Cáceres, R., Marfà, O. 2016. The effect of precomposted sewage sludge mixture amended with biochar on the growth and reproduction of Eisenia fetida during laboratory vermicomposting. *Ecological Engineering*, 90, 35–41.

Mannina, G., Presti, D., Montiel-Jarillo, G., Carrera, J., Suárez-Ojeda . M.E. 2020. Recovery of polyhydroxyalkanoates (PHAs) from wastewater: A review. *Bioresource Technology*, 297, 122478.

Maruri-López, I., Aviles-Baltazar, N.Y., Buchala, A., Serrano, M. 2019. Intra and extracellular journey of the phytohormone salicylic acid. *Frontiers in Plant Science*, 10, 423.

Mata-Alvarez, J. 2003. *Biomethanization of the Organic Fraction of Municipal Solid Waste*. IWA Publshing, London.

Meegoda, J.N., Li, B., Patel, K. and Wang, L.B. 2018. A review of the processes, parameters, and optimization of anaerobic digestion. *International Journal of Environmental Research and Public Health*, 15, 10.

Monroy, F., Aira, M., Domínguez, J. 2009. Reduction of total coliform numbers during vermicomposting is caused by short-term direct effects of earthworms on microorganisms and depends on the dose of application of pig slurry. *Science of the Total Environment*, 407(20), 5411–5416.

Mukherjee, A., Zaveri, P., Patel, R., Shah, M.T., Munshi, N.S. 2021. Optimization of microbial fuel cell process using a novel consortium for aromatic hydrocarbon bioremediation and bioelectricity generation. *Journal of Environmental Management*, 298, 113546.

Mulla, S.I., Ameen, F., Talwar, M.P., Eqani, S.A.M.A.S., Bharagava, R.N., Saxena, G., Tallur, P.N., Ninnekar, H.Z. 2020. Organophosphate pesticides: Impact on environment, toxicity, and their degradation. In: G. Saxena, R. Bharagava (eds) *Bioremediation of Industrial Waste for Environmental Safety*. Springer, New York, pp. 265–290.

Nadana, G.R.V., Rajesh, C., Kavitha, A., Sivakumar, P., Sridevi, G., Palanichelvam, K. 2020. Induction of growth and defense mechanism in rice plants towards fungal pathogen by eco-friendly coelomic fluid of earthworm. *Environmental Technology & Innovation*, 19, 101011.

Naowasarn, S., Leungprasert, S. 2016. Bioremediation of oil-contaminated soil using chicken manure. *Soil and Sediment Contamination: An International Journal*, 25(7), 739–756.

Pandey, G., Bajpai, S., Tripathi, S., Chauhan, R., Sharma, P. 2022. Approaches, challenges, and prospects of life cycle assessment applied in nano bioremediation of contaminated sites. *Nanotechnology*, 15(2), 1108–1125.

Park, M.J., Kim, J.H., Lee, Y.H., Kim, H.M., Jeong, D.W. 2020. System optimization for effective hydrogen production via anaerobic digestion and biogas steam reforming. *International Journal of Hydrogen Energy*, 45(55), 30188–30200. doi:10.1016/j.ijhydene.2020.08.027.

Parthipan, P., Cheng, L., Dhandapani, P., Elumalai, P., Huang, M., Rajasekar, A. 2022. Impact of biosurfactant and iron nanoparticles on biodegradation of polyaromatic hydrocarbons (PAHs). *Environmental Pollution*, 306, 119384.

Patel, A.K., Singhania, R.R., Albarico, F.P.J.B., Pandey, A., Chen, C.-W., Dong, C.-D. 2022. Organic wastes bioremediation and its changing prospects. *Science of the Total Environment*, 824, 153889.

Peng, Y., Lu, S., Li, X., Yan, J., Cen, K. 2020. Formation, measurement, and control of dioxins from the incineration of municipal solid wastes: Recent advances and perspectives. *Energy & Fuels*, 34(11), 13247–13267.

Procházková, P., Hanč, A., Dvořák, J., Roubalová, R., Drešlová, M., Částková, T., Šustr, V., Škanta, F., Pacheco, N.I.N., Bilej, M. 2018. Contribution of Eisenia andrei earthworms in pathogen reduction during vermicomposting. *Environmental Science and Pollution Research*, 25(26), 26267–26278.

Quintella, C.M., Mata, A.M.T., Lima, L.C.P. 2019. Overview of bioremediation with technology assessment and emphasis on fungal bioremediation of oil contaminated soils. *Journal of Environmental Management*, 241, 156–166.

Quintern, M., Morley, M. 2017. Vermicomposting of biosolids and beneficial reuse-New Zealand commercial case studies from 4 communities over 8 years. In: *Residuals and Biosolids Conference 2017*. Water Environment Federation, Alexandria, VA.

Ragoobur, D., Huerta-Lwanga, E., Somaroo, G.D. 2022. Reduction of microplastics in sewage sludge by vermicomposting. *Chemical Engineering Journal*, 450, 138231.

Rajendran, N., Jeehoon, H. 2022. Techno-economic analysis of food waste valorization for integrated production of polyhydroxyalkanoates and biofuels. *Bioresource Technology*, 348, 126796.

Ramezani, M., Rad, F.A., Ghahari, S., Ghahari, S., Ramezani, M. 2021. Nano-bioremediation application for environment contamination by microorganism. In: D.G. Panpatte, Y.K. Jhala (eds), *Microbial Rejuvenation of Polluted* Environment. Springer, New York, pp. 349–378.

Rapport, J.L., Zhang, R., Williams, R.B., Jenkins, B. 2012. Anaerobic digestion technologies for the treatment of municipal solid waste. *International Journal of Environment and Waste Management*, 9(1–2), 100–122. doi:10.1504/IJEWM.2012.044163

Ritchie, H., and Roser, M. 2018. Plastic Pollution. Our World in Data, Oxford.

Rosenboom, Jan-Georg, Langer, R., Traverso, G. 2022. Bioplastics for a circular economy. *Nature Reviews Materials*, 7(2), 117–137.

Romio, C., Kofoed, M.V.W., Møller, H.B. (2021). Digestate post-treatment strategies for additional biogas recovery: A review. *Sustainability*, 13(16), 9295. doi:10.3390/su13169295

Roubalová, R., Procházková, P., Hanč, A., Dvořák, J., Bilej, M. 2020. Mutual interactions of E. andrei earthworm and pathogens during the process of vermicomposting. *Environmental Science and Pollution Research*, 27(27), 33429–33437.

Saha, S., Mina, B.L., Gopinath, K., Kundu, S., Gupta, H. 2008. Relative changes in phosphatase activities as influenced by source and application rate of organic composts in field crops. *Bioresource Technology*, 99(6), 1750–1757.

Sahni, S., Kumar, S., Prasad, B. 2021. Integration of salicylic acid, vermicompost and bioagent for effective management of chickpea wilt disease. *Journal of Environmental Biology*, 42(5), 1274–12280.

Sahota, S., Shah, G., Ghosh, P., Kapoor, R., Sengupta, S., Singh, P., Vijay, V., Sahay, A., Vijay, V.K., Thakur, I.S. 2018. Review of trends in biogas upgradation technologies and future perspectives, *Bioresource Technology Reports*, 1, 79–88. doi:10.1016/j.biteb.2018.01.002.

Saratale, R.G., Cho, S.-K., Saratale, G.D., Kadam, A.A., Ghodake, G.S., Kumar, M., Bharagava, R.N., Kumar, G., Kim, D.S., Mulla, S.I. 2021. A comprehensive overview and recent advances on polyhydroxyalkanoates (PHA) production using various organic waste streams. *Bioresource Technology*, 325, 124685.

Saxena, G., Kishor, R., Bharagava, R.N., Das, P., Gupta, P.K., Kumar, N. 2021. Emerging green technologies for biological treatment of leather tannery chemicals and wastewater. In: G. Saxena, V. Kumar, M.P. Shah (eds), *Bioremediation for Environmental Sustainability*. Elsevier, Amsterdam, pp. 435–457.

Sebastian, P.M., Rao, K. 2022. Approaches in bioremediation of dioxins and dioxin-like compounds-a review on current and future prospects. *Nature Environment & Pollution Technology*, 21, 3.

Sehrawat, A., Phour, M., Kumar, R., Sindhu, S.S. 2021. Bioremediation of pesticides: An eco-friendly approach for environment sustainability. In: D.G. Panpatte, Y.K. Jhala (eds), *Microbial Rejuvenation of Polluted* Environment. Springer, New York, pp. 23–84.

Sfetsas, T., Patsatzis, S., Chioti, A. et al. (2022). A review of advances in valorization and post-treatment of anaerobic digestion liquid fraction effluent. *Waste Management & Research*, 40(8), 1093–1109. doi:10.1177/0734242X211073000

Shi, J., Wang, Z., Stiverson, J.A., Yu, Z., Li, Y. 2013. Reactor performance and microbial community dynamics during solid-state anaerobic digestion of corn stover at mesophilic and thermophilic conditions. *Bioresource Technology*, 136, 574–581. doi:10.1016/j.biortech.2013.02.073.

Shi, Z., Liu, J., Tang, Z., Zhao, Y., Wang, C. 2020. Vermiremediation of organically contaminated soils: Concepts, current status, and future perspectives. *Applied Soil Ecology*, 147, 103377.

Sillero, L., Solera, R., Perez, M. 2022. Anaerobic co-digestion of sewage sludge, wine vinasse and poultry manure for bio-hydrogen production. *International Journal of Hydrogen Energy*, 47(6), 3667–3678, doi:10.1016/j.ijhydene.2021.11.032.

Singh, R., Soni, S., Awasthi, A., Kalra, A. 2012. Evaluation of vermicompost doses for management of root-rot disease complex in Coleus forskohlii under organic field conditions. *Australasian Plant Pathology*, 41(4), 397–403.

Statista. 2022a. Biogas energy capacity worldwide from 2009 to 2021 in megawatts. https://www.statista.com/statistics/1032922/biogas-capacity-globally/#:~:text=Installed%20capacity%20of%20biogas%20energy%20worldwide%20reached%20some%2021.4%20gigawatts%20in%202021

Statista. 2022b. Production of biogas worldwide in 2019, by region in billion m3. https://www.statista.com/statistics/481828/biogas-production-worldwide-by-region/

Talan, A., Kaur, R., Tyagi, R.D., Drogui, P. 2020. Bioconversion of oily waste to polyhydroxyalkanoates: Sustainable technology with circular bioeconomy approach and multidimensional impacts. *Bioresource Technology Reports*, 11, 100496.

TotalEnergies. 2023. PLA: Becoming one of the leaders in the bioplastics market. https://totalenergies.com/energy-expertise/projects/bioenergies/pla-bioplastic

Trchounian, A. 2015. Mechanisms for hydrogen production by different bacteria during mixed-acid and photo-fermentation and perspectives of hydrogen production biotechnology. *Critical Reviews in Biotechnology*, 35(1), 103–113.

US EPA. 2022. Environmental benefits of anaerobic digestion (AD). https://www.epa.gov/anaerobic-digestion/environmental-benefits-anaerobic-digestion-ad

US EPA. 2016. Environmental Benefits of Anaerobic Digestion. https://www.epa.gov/anaerobic-digestion/environmental-benefits-anaerobic-digestion-ad.

USDA. 2023. Biobased products. https://www.biopreferred.gov/BioPreferred/faces/pages/BiobasedProducts.xhtml, accessed January 12, 2023.

Usmani, Z., Kumar, V., Gupta, P., Gupta, G., Rani, R., Chandra, A. 2019. Enhanced soil fertility, plant growth promotion and microbial enzymatic activities of vermicomposted fly ash. *Scientific Reports*, 9(1), 1–16.

Varjani, S.J. 2017. Microbial degradation of petroleum hydrocarbons. *Bioresource Technology*, 223, 277–286.

Verma, J., Kumar, D., Singh, N., Katti, S.S., Shah, Y.T. 2021. Electricigens and microbial fuel cells for bioremediation and bioenergy production: A review. *Environmental Chemistry Letters*, 19(3), 2091–2126.

Vicentin, R., Masín, C.E., Lescano, M.R., Zalazar, C.S. 2021. Poultry litter stabilization by two-stage composting-vermicomposting process: Environmental, energetic and economic performance. *Chemosphere*, 281, 130872.

Vögeli, Y., Lohri, C.R., Gallardo, A., Diener, S., Zurbrügg, C. 2014. *Anaerobic Digestion of Biowaste in Developing Countries: Practical Information and Case Studies*. Swiss Federal Institute of Aquatic Science and Technology (Eawag), Dübendorf, Switzerland

Wang, Z., Liu, T., Duan, H., Song, Y., Lu, X., Hu, S., Yuan, Z., Batstone, D., Zheng, M. 2021. Post-treatment options for anaerobically digested sludge: Current status and future prospect, *Water Research*, 205. doi:10.1016/j.watres.2021.117665.

Wellinger, A., Murphy, J.P., Baxter, D. (Eds.). 2013. *The Biogas Handbook: Science, Production and Applications*. Elsevier, Amsterdam.

Wu, Y., Wang, S., Cheng, F., Guo, P., Guo, S. 2020. Enhancement of electrokinetic-bioremediation by ryegrass: Sustainability of electrokinetic effect and improvement of n-hexadecane degradation. *Environmental Research*, 188, 109717.

Xia, H., Wu, Y., Chen, X., Huang, K., Chen, J. 2019. Effects of antibiotic residuals in dewatered sludge on the behavior of ammonia oxidizers during vermicomposting maturation process. *Chemosphere*, 218, 810–817.

Yadav, B., Pandey, A., Kumar, L.R., Tyagi, R.D. 2020. Bioconversion of waste (water)/residues to bioplastics-A circular bioeconomy approach. *Bioresource Technology*, 298, 122584.

Yakkou, L., Houida, S., Domínguez, J., Raouane, M., Amghar, S., El Harti, A. 2021. Identification and characterization of microbial community in the coelomic fluid of earthworm (Aporrectodea molleri). *Microbiology and Biotechnology Letters*, 49(3), 391–402.

Yuvaraj, A., Thangaraj, R., Ravindran, B., Chang, S.W., Karmegam, N. 2021. Centrality of cattle solid wastes in vermicomposting technology-A cleaner resource recovery and biowaste recycling option for agricultural and environmental sustainability. *Environmental Pollution*, 268, 115688.

Zhu, C., Chiu, S., Nakas, J.P., Nomura, C.T. 2013. Bioplastics from waste glycerol derived from biodiesel industry. *Journal of Applied Polymer Science*, 130(1), 1–13.

# 2 Phages and Their Role in Natural and Directed Microbial Control

*Amro Abd Al Fattah Amara*

## 2.1 INTRODUCTION

The natural components (physical, chemical, and biological) are highly dynamic and stable. Nature has self-solutions for its own problems. Recognizing and analyzing the different natural mechanisms are important. Nature is able to readjust spontaneously any deviations in its ecosystem balance. Some of the natural activities are raised by emergency actions; for example, the natural balance between prokaryote cells and the phages. Understanding such a balance will enable the design of better solutions for our own-made problems. That will enable us to manage different types of wastes. Many environmental biophysical factors including virophages, nutrients (macronutrients and micronutrients), temperature, seasonal variations, light, solar UV radiation, pH, cation concentrations, salinity, and microbial community are affecting the phages activities (Jassim and Limoges, 2013; Jassim and Limoges, 2014; Jassim and Limoges, 2017). Recognizing these factors is essential for better use of the phages in various biological applications (Abdulamir et al., 2014; Jassim et al., 2012; Jassim and Limoges, 2014).

Additional phage features enable solving some critical problems (described below). That becomes available after the progress in the field of bioengineering, biotechnology, molecular microbiology, etc. In most cases, these phage applications mimic natural activity, or introduce *de nevo* applications and *de novo* strategies. The processes could include using the wastes as substrates. Such a bio-cycle degrades a waste and produces a product (e.g. *Spirulina platensis*). In nature, the lysed microbes are part of the prokaryotic and eukaryotic food cycle. Phages are working collectively as hidden soldiers against many harmful microbes. The research concerning the bacteriophage and their applications is a hot subject. One report describes that more than 5,100 bacteriophages have been identified and studied.

Ninety percent of phages possess tails. They belong to the Myoviridae, Siphoviridae, and Podoviridae families (Ackermann, 2006; Wittebole et al., 2013). The DNA of the bacteriophage has been used by Hershey and Chase and has been proved that DNA is the code of the genetic material by labeling its precursors and investigating its presence in the new phages. Many finished bacteriophage genomes can be found in the NCBI Genome database. All of that proves the fact that phages are everywhere. Human hopes have been concerned with applying the phages to control different microbes. Limitations in phage usage led to neglecting using phages as a biocontrol agent. Unfortunately, phages can cause horizontal gene transfer within microbes.

After the emergence of the antibiotic resistance, phages have come into focus, yet again. They are suggested as a powerful alternative to antibiotics, particularly in superficial treatments. Bacteriophages give a positive sign as complementary to antibiotics. Isolated bacteriophages show *in vitro* ability to lyse clinically isolated MRSA strains. The used *Staphylococcus aureus* strains contain heterogeneous vancomycin and teicoplanin-resistant strains (O'Flaherty et al., 2005).

DOI: 10.1201/9781003386902-3

Sharaf et al. (2018) have reported the successful survival of *Arthrospira fusiformis* for several years in a flask. That has been explained because the existence of complete food cycle includes the nutrients, sunlight, *A. fusiformis*, bacteria, and phages. *A. fusiformis* use sunlight in the photosynthesis, and the bacteria degrade their dead cells. While the bacteriophage reduces the bacterial number and lyses them, the lysis process provides cells (*A. fusiformis*) with nutrients (Sharaf et al., 2018).

Microbial lysis by phages occurs in nature. The lysed cells' cytoplasm will escape to the surrounding media. It will supply other biological structures with necessary compounds. These food cycles can be controlled and used to flourish the production of certain edible microbes. In addition, they can be designed as a system for waste degradation/bacterial lysis/algae production. In fact, all of these have happened in nature spontaneously.

## 2.2 HISTORY OF THE PHAGE DISCOVERY

Ernest Hankin has written early reports about the activity of bacteriophages in 1896. A report in 1915 by Twort described a phenomenon, called "glassy transformation" (of bacterial colonies on agar) and "transmissible lysis," that has been caused by bacteriophage in its cultures (d'Herelle, 1917; Goodridge and Abedon, 2003; Summers, 2016; Twort, 1915). In 1917, d'Herelle published a brief note in the Comptes rendus de l'Academie des sciences in which he described a new kind of microbe, in his words, "an obligate intracellular parasite" of bacteria. In 1919, d'Herelle used phages to treat dysentery and developed a commercial laboratory in Paris that produced phage. The lab became the large French company L'Oréal. In the 1930s, therapeutic phages were marketed in the United States by companies including Eli Lilly, Squibb & Sons (today Bristol-Meyers Squibb), and the Swan-Meyers division of Abbott Laboratories.

The success of the antibiotics as potent antimicrobial agents was affected negatively on research concerning the bacteriophages, in most of the Western world. Meanwhile, phage research has been further developed in Eastern Europe and the former Soviet Union. Some centers have been active such as the Eliava Institute of Bacteriophage, Microbiology, and Virology (EIBMV) in Tbilisi, Georgia. Other institutes are in Russia, and the Hirszfeld Institute of Immunology and Experimental Therapy in Wroclaw, Poland (Pirnay et al., 2010). In France, therapeutic phage preparations (made-to-order from) in the Institut Pasteur (Paris and Lyon) have been used until the beginning of the nineties. Commercial phage preparations are used to treat severe infections (Verbeken et al., 2007),

Phages to treat burn wounds have been approved in Belgium (Kutter et al., 2010). Nowadays, phages are commercially available under different trade names such as ListShield™, EcoShield™, Agriphage™, Listex™, and Salmonellex™ (Madhusudana Rao and Lalitha, 2015).

## 2.3 THE PHAGES

### 2.3.1 The Phage Feature

Under the molecular biology and the genetic level, phages have been studied early by Alfred Hershey, Max Delbruck, and Salvador Luria who discovered that viruses could exchange genetic material. In 1969, the "phage group" (Hershey, Delbruck and Luria) was awarded the Nobel Prize for Physiology or Medicine for their discoveries concerning "the replication mechanism and the genetic structure of viruses." Like all viruses, phages are simple but unique organisms with a core of genetic material (nucleic acid) surrounded by a protein capsid. They could infect prokaryotic cells (bacteria, cyanobacteria, and archaea). Meanwhile different prokaryotic structures are resistant naturally to the phages or they acquired resistance agents.

When phages infect bacteria, they are named bacteriophages or bacterial viruses. If they infect other prokaryotic structures, they are named phages or take the first name of their host like cyanophage that could infect cyanobacteria. Phages have complex structures but generally have a head and complex tail structures. The head contains its genomic elements. A few phages have lipids in their envelope; however, most of them do not have them. They are either very small or big phages with head and tail. The tail is involved in injecting nucleic acid into the cell. The tail is a variant (big or small) based on the type of the phage. On Petri dishes, they show different lysis structures for the bacteria (different clear zones). Bacteriophages are well studied for some bacterial group such as *E. coli* and *Salmonella typhimurium* which belong to the enterobacteria.

### 2.3.2 Phages as Prokaryotic Cell Dependent

Bacteriophages are biological structures that are inactive outside their hosts. Therefore, they do not have any metabolic machinery that possesses affinity to eukaryotic cells (Chan and Abedon, 2012; Dabrowska et al., 2005).

One unique criterion of the phages that attracts attention early is their host specificity. Phages have intrinsic metabolism if present in the host cytoplasm (Withey et al., 2005). They control the host metabolic machinery and direct it for their own production and reproduction. After they enter the bacterial cells, they control their genetic materials and start to reproduce to generate their new progeny.

The process, which starts from the settlement of the phage on the bacterial surface, and injecting their genetic material into the bacterial cytoplasm, ends by rupturing the bacterial cells. This step is essential for releasing the new progeny to the outer environment. The new progeny is ready to infect other bacterial cells. In fact, before such an understanding and before the release of the science of molecular biology, the phages are characterized by their host and by their lysis-imaging feature. In contrast to antibiotics, bacteriophages are mostly very selective with regard to the targeted pathogens.

Besides being selective with regard to the prokaryotic cells (not infecting the eukaryotic cells), phages could be very specific on the species or strain level. In addition, bacteriophages used for treating zoonotic pathogens have short generation time, are nontoxic, very specific, and safe for the other microflora microbes (Hudson et al., 2013; Rong et al., 2014).

### 2.3.3 Phage Infection Mechanism and Purification

The phages infect the prokaryotic cells through passive diffusion. Based on their type, they are either very specific to their host or nonspecific. Their adsorption and entry have happened through specific receptors on the cell surfaces (e.g., protein, carbohydrate and lipopolysaccharides (LPS)) (Marks and Sharp, 2000). The phage-receptor influences the bacterial host range, which is narrow for aquatic phages and broad (polyvalent) in cyanophages (Suttle, 2000) and sewage.

The DNA bacteriophages infect the enteric group but are specific only to those with pili (gene donors). When MS2 (RNA bacteriophage) injects its RNA in the host ribosome, four genes are translated: the maturation protein, coat protein, lysis protein, and RNA replication. Phage assembly occurs and is released through various cell lysis. Brüssow and Kutter (2005) reported that bacteriophage numbers in the environment are ten times more than bacteria (Brüssow and Kutter, 2005). The bacteriophages are generally classified based on their genomic materials into four groups: dsDNA, ssDNA, dsRNA, and ssRNA (Ackermann, 2006). Meanwhile dsDNA bacteriophages are the abundant group (Myoviridae, Podoviridae and Siphoviridae), particularly the order of Caudovirales (Wittebole et al., 2013).

Scientists have observed this phenomenon, which could happen particularly during cultivating microbes on agar plates (e.g., from water samples). The bacteriophages can be observed only through the electron microscope. Bacteriophages' activity can be spotted as a clear zone because of the death of the bacteria. This natural phenomenon has been well investigated and bacteriophage isolation protocols are well adapted. Interestingly, many microbiologists start working with these bacterial viruses because of unexpected phage attacks on their bacterial cells. For isolating specific phages, it should be isolated from the ecosystem where the specific bacteria or other microbes exist. Alternatively, an enriched strategy by the targeted microbe using the expected source that contains the specific phage will lead to the isolation of the proper phage for the used microbe. Traditionally, bacteriophages have been purified by polyethylene glycol (PEG) precipitation (Han et al., 2014) or cesium chloride (CsCl) gradients centrifugation (Wong et al., 2014). They can be characterized in terms of morphology, genomic DNA, and proteomic and biological measurements of phage host infections (Sun et al., 2012; Tiwari et al., 2013; Yang et al., 2010).

Bacteriophages are everywhere in the nature, particularly in the soil, oceans, fresh waters, and in our food. As mentioned above, phages are diverse in their structure, hosts, and environment as well. However, one should note that they are mainly searching for their host, the prokaryotic microbes that are already diverse in their nature. In this regard, it is a fact that different bacteriophages can be isolated from different ecosystems.

### 2.3.4 Lytic and Lysogenic Cycle

The normal phage cycle, which ends with the hosts' cell lysis, is named the lytic cycle (virulent). The number of progenies range from 50 to 200 (Wommack and Colwell, 2000) Another type of infection is named the lysogenic (temperate) infection, wherein the phage nucleic acid recombines with the host cell genome. The new insert is named a dormant prophage. It reproduces with the cell genome and protects the cell from phage infections.

Changes in the environment, usually those causing stresses, such as UV light and chemical mutagens, can activate the dormant prophage to start the lytic cycle (Jiang and Paul, 1998). Once the lysogenic cycle is adopted, the phage chromosome is integrated (by covalent bonds) into the bacterial chromosome, where it can remain dormant for thousands of generations. The lysogenic phage transfer virulence factors are inherited from previously infected bacteria via transduction (Campbell, 1988). The phage and the bacteria are in action and interact with each other and with different environmental effects (Jassim and Limoges, 2014; Maxwell, 2016). Apparently, because of their lysogenic cycle, Jassim and Limoges (2013, 2014) suggested that it might be better to consider phages as opportunistic living biological agents (Jassim and Limoges, 2013, 2014). The phages can store part of the bacterial DNA in their capsid (which encloses the genetic material). During the next infection cycle, the bacterial host might integrate the foreign DNA by homologous recombination to the bacterial DNA (Zhang et al., 2011).

Evidence shows that broad-host-range plasmids have been transduced between bacteria of non-related species (Evans et al., 2010). For example, it has been found that phages can transduce resistance genes from *Salmonella enterica* serovar *typhimurium* DT104 (Schmieger and Schicklmaier, 1999). The temperate phages are problematic because of converts turning phage-sensitive bacteria into insensitive ones, and the encoding of bacterial virulence factors, including bacterial toxins adhesion, hemolysin, serine protease, thermostable enterotoxin, and special catalase system (Alexa et al., 2001; Hayashi, 2001; Kutter et al., 2010; Vondruskova et al., 2010).

## 2.4 PHAGE APPLICATIONS

Phages have many applications, including research, medicine, food applications, plant and animal gene transfer, phage display–based selection for biological affinity molecules, bacterial biosensing devices, genes delivery, food biopreservation and safety, teeth decay control, biocontrol of plant pathogens, biofilm control, surface disinfection, corrosion control, wastewater treatment, molecular biology protocols concerning nucleic acid replications, protein synthesis, bacterial host, phage display preparation, protein overexpression (e.g. T7 promotor). They also act as a biopreservative helping in biocontrol (Bardina et al., 2012; Heringa et al., 2010; Jaiswal et al., 2013; Larbanoix et al., 2011; Liu et al., 2015a; Lone et al., 2016; Zinno et al., 2014) and deliver reporter genes (e.g. lux) (Schmelcher and Loessner, 2014). Bacteriophage particles have been used with a fluorescent dye covalently attached to their protein capsids for detecting specific adsorption (Goodridge et al., 1999). In addition, they also detect released cellular intracytoplasmic components (e.g., adenylate kinase). Selective applications will be included in this chapter.

### 2.4.1 Phages as Therapeutic Agents

The phages' unique criteria enable their integration as therapeutic agents, particularly as an alternative to antibiotics. Early trials to treat infectious diseases by bacteriophage such as bubonic plague and cholera have been examined. Applying bacteriophages to treat and control bacterial infections (e.g., pathogenic bacteria) is considered since the antibiotics are widely commercialized. Another purpose is their ability to develop horizontal gene transfer, though it does not apply for all phages. Some other conditions are negatively effect on the phages' activity such as their inability to attack bacteria in viscous solution such as the injury's bus.

Few diagnostic kits have reached the market for pathogen detection in human samples (e.g., FastPlaqueTB assay (BIOTEC Laboratories Ltd, *Mycobacterium tuberculosis*), KeyPath test (MicroPhage Inc.) *Staphylococcus aureus* MRSA/MSSA, *S. typhimurium*, and MRSA/MSSA (Blood Culture Test) (Henry and Debarbieux, 2012; Schmelcher and Loessner, 2014; Sorokulova et al., 2014). Phage has been used in different typing schemes for pathogenic bacteria detection. This technique can detect only viable cells and is used particularly in identifying *Campylobacter*, *Escherichia coli*, *Listeria*, *M. tuberculosis*, *Pseudomonas*, and *Salmonella* (Oliveira et al., 2015; Pai et al., 2005; Stewart et al., 1998).

The capacity to genetically engineer phages with the new therapeutic properties is potentially advantageous. Bacteriophages are effective if they directly interact and kill disease-causing bacteria. They must be administered around the time of infection. They will work only when bacteria are actually present in abundance and cause infections. The absence of bacteria may be rapidly inactivated in the animal. Some components of phages are used successfully at the industrial level to enable lysis of the producer strains. Phages have *E* lysis genes. They have been cloned and used to induce microbial cell lysis (Amara, 2016; Laemmli, 1970; Panthel et al., 2003) (and the references within). After specific bacteriophage-mediated lysis, bacteriophage particles displaying peptides or antibody fragments that specifically bind bacterial pathogens or toxins are used (Petrenko and Vodyanoy, 2003).

Using a bacteriophage display technique, phages have been engineered to express a biotin-binding domain on a capsid protein. The chemical attachment of genetically biotinylated T4 bacteriophage particles onto streptavidin-coated gold surfaces (Gervais et al., 2007) or streptavidin-coated magnetic beads and microcrystalline cellulose beads has also been reported. For that, both of the complete phages or their genes are used in different therapeutic applications. Bacteriophage endolysins cover a diverse range of hydrolytic enzymes. Most of them are species-specific. Endolysins contain an enzymatically active and cell wall–binding domain (Roach and Donovan, 2015).

The first commercially available endolysin, Staphefekt (Micreos Human Health, The Netherlands), is available for the treatment of human skin infections caused by *S. aureus*. Lysins are phage-encoded peptidoglycan hydrolases. It is prepared as purified recombinant proteins. It is able to lysis the Gram-positive bacteria if applied exogenously (Young, 1992; Fischetti, 2008).

Bacteriophages have been formulated in different applications as liquids and creams, and impregnated into solids, capsules, and tablets (Abdulamir et al., 2014; Kutter et al., 2010). Phage cocktail formulation and dosage should be optimized to more precisely kill pathogenic microorganisms. The formulated virus/phage with fluorescent brighteners, skim milk alone or in combination with sucrose, has been shown to enhance their activity under intense UV irradiation (Balogh et al., 2003).

During the phage formulation, their concentrations should be maintained at the proper level. Phage concentration or number should be maintained to enable a collective bacterial lysis rate, which is higher than the rate of the collective bacterial multiplications. Another important point in the directed phage treatment is to maintain the phage number in a certain area. One method is to encapsulate the phage in liposomes (Mathieu et al., 2019). This prevents the phage from depleting before reaching the target bacteria.

Microfluidics is useful for encapsulating phages of various sizes into liposomes with reasonable efficiency and consistent titers (Mathieu et al., 2019) Another study used bacterial endospores as a protective cover for phages. Endospores show superior tolerance to harsh environments such as elevated temperature and extreme pH compared to free-living phages (Gabiatti et al., 2018).

One major problem in the industrial sector that depends on microbial fermentation such as antibiotic production is the loss of the process because of a sudden attack by the bacteriophage. Of course, the losses are mainly because of the reduction in the bacterial cell number because of the bacteriophage's activity and the cost of the effort to cure the fermentors and the industrial area from such an unpleasant visitor.

Like the antibiotics, the bacteriophages could not guarantee a perfect full control of the microbes. In addition, there is a need to use them wisely in the other applications, particularly in the open nature applications where the control might not be available. Meanwhile, if certain bacteriophage types do not naturally exist in certain ecosystem, it should be used under control to avoid side effect to the nature.

### 2.4.2 Phage as Biocontrol Agent in the Food Industry

Bacteriophages can be used as food biopreservatives (Henry and Debarbieux, 2012; Tan et al., 2014). The bacteriophage potential as biocontrol agents in the food industry shows rising focus and research interest. OmniLytics Inc. (Sandy, UT, United States) and Intralytix Inc. (Baltimore, MD, United States) have developed different bacteriophages against *Salmonella*, *E. coli*, and *Listeria monocytogenes* in the food industry. In Europe, a Dutch company Micreos BV (Wageningen, Netherlands) marketed bacteriophage products against *Salmonella* and *E. coli*. A German company, Fink Tec (Hamm, Germany), targets *E. coli* (Doss et al., 2017; Moye et al., 2018). Specific types of bacteriophages can reduce the load of *Campylobacter*, *Listeria*, *Salmonella*, and others in different foods (Hudson et al., 2005; Lone et al., 2016; Oliveira et al., 2015).

*Salmonella* bacteriophages are used by dipping or spraying or as a liquid to different food matrices (Bao et al., 2015; Spricigo et al., 2013; Sukumaran et al., 2015; Zinno et al., 2014). In poultry industries, endolysin Ply3626 has been found to be active against *Clostridium perfringens* (Zimmer et al., 2002).

### 2.4.3 Control Cyanobacteria Bloom-Forming by Cyanophages

Harmful cyanobacterial algal blooms have different negative impacts such as taste-and-odor problems in drinking water and aquaculture resources, animal deaths, and reduced recreational opportunities and tourism. Cyanophages have high lytic efficiency. Therefore, cyanophages have been isolated, including PaV-LD and MaLMM01 (Gao et al., 2012; Yoshida-Takashima et al., 2012). *nblA*, a key gene for phycobilisome degradation, has been found in several cyanophages (Gao et al., 2012; Yoshida-Takashima et al., 2012). These cyanophages can deplete phycobilisomal proteins and disrupt the thylakoid form of cyanobacteria. They are used in manipulating dangerous buds. A possible solution for cyanophages resistance is by mixing a high concentration of cyanophages with a buffer in special capsules that could release them under pressure. They have greater lytic ability and can control resistant cyanobacteria (Jassim and Limoges, 2017). However, these approaches might be affected by the changes in the water currents and the unfavorable weather conditions.

It is important to highlight that many of them are host-specific and are capable of entering into a lysogenic relationship with their hosts under the influence of external forces in aquatic ecosystems, direct sunlight, phosphate concentrations, $CO_2$, ozone and virophage (Jassim and Limoges, 2013). Therefore, controlling such factors that enhance the lysogenic cycle will facilitate the phage activity. For example, solar protecting devices have been identified to increase biocontrol efficacy not only for phages but also for entomopathogenic viruses and proteinaceous biopesticides (Behle et al., 1996; Ignoffo et al., 1997).

### 2.4.4 Treatment of Bacteria Contaminated Biofilms

Biofilms consist of a diverse group of bacteria that can synergistically resist invasion by antimicrobial agents and other bacteria (Pires et al., 2017). Different mechanisms can be involved in reducing the attack on the biofilm microbes by the antimicrobial agents, particularly the antibiotics. A microbial biofilm causes a continuous decrease in the net waste flow at the membrane surface, which significantly affects the efficacy of the device (Wu et al., 2017). For example, a biofilm formed by *Delftia tsuruhatensis* has been isolated from a sewage treatment system on membrane filters treated with phage lytic DTP1 (Bhattacharjee et al., 2015). The membrane flux increased. *E. coli* phage P2 has been used to purify ARB-infected nanocomposite membranes. The phage and modified nanocomposite membrane combination solved the problems of bacterial and protein entrapment and increased membrane permeability (Ayyaru et al., 2018). Using electron microscopy enables the detecting of phages and prevents the formation of biofilms on the surface of the membrane.

### 2.4.5 Water, Wastewater, and Sludge Treatment

There are different sources responsible for the pollution of the drinking water. The people's need for clean drinking water is rising daily, but the world's water supply is limited. It is estimated that about one-ninth of the world's inhabitants do not have access to safe water. Even in countries that have great rivers, pure safe water might not be available. Proper wastewater treatment and recycling are becoming important to address this imbalance. Harmful pathogens in wastewater negatively affect human health, particularly if drinking water is contaminated (Ji et al., 2021; Mathieu et al., 2019). For example, pharmaceutical residues from hospitals can enter wastewater from different sources such as the pharmaceutical companies' wastes, hospitals, animal farming discharge, public consumption and the like, and finally could reach drinking water (Deng et al., 2018; Mathieu et al., 2019; Zhang et al., 2019).

Microbes that existed in certain native ecosystem are affected if polluted water contaminate it. For that, the treatment of the polluted water is an essential step. Such pressure (due to polluted drinking water) could affect a certain group of bacteria (Khan and Gupta, 2020). They can be eradicated or their number could be reduced. This will leave the other group (that are resistant naturally) to be dominant. Mostly, they are harsh microbes that could resist different killing factors. For that, one needs to use selective system that could control any pathogens. Different alternative mechanisms can be employed such as accelerating the transfer of the resistance virulence genes or mutating one to be resistant to another type of antibiotic. Such mechanisms collectively collaborate to accelerate antibiotic resistance development (Deng et al., 2018; Mathieu et al., 2019; Zhang et al., 2019). For that, it is wise to investigate water before treating so that it can be treated specifically and selectively. Any chemical compounds that enter the water kill harmful microbes but also spoil our health and reduce our chance of drinking pure safe water. Bacteriophage has been used to detect the presence of its bacterial host. For any treatment process, it is logical to use phage selectively. Interestingly, the phages can be used to map the existing pathogenic bacteria in water. Phage indicators can be divided into three classes based on the bacterial strain that can cause illness and contaminate water from the human fecal source.

*E. coli* phages, *Bacteroides* species phage, and enterococcal phage. Somatic coliphage (SOMPH) and FRNA coliphage are two types of *E. coli* phage (F-RNAPH). It is also logical to isolate the specific phage from its source directly, propagate it to the accepted concentration, and reuse it in the treatment. One can isolate existing phages from water sources contaminated by bacterial strains or enrich the water by the targeted strain (in lab) and investigate the effect of different existing phages (each one alone). For example, three human-specific phages that infect *Bacteroides fragilis* have been traced: (Bf124PH), F-RNAPH, and SOMPH. SOMPH had a high detection rate and a high concentration, and has been well tolerated (Dias et al., 2018). A meta-analysis found that the average LRV of different phages (SOMPH, F-RNAPH, MS2 coliphage, and T4 phage) differ slightly. The cross-aggregating phages (crAssphage) are commonly found in the human gut metagenome. crAssphage (FCrAss001) has been recovered from human feces and identified as a *Bacteroides intestinalis* phage (Shkoporov et al., 2018). Global primary phage-based generation further reduces the cost because of using different chemical reagents. Isolation of specific phage against specific pathogen is useful in protecting against infections (Mathieu et al., 2019). FAB2 phage isolated from sanatorium wastewater can infect different *A. baumannii* strains (Lin et al., 2010). Phages are good tools for restricting the *Vibrio cholerae* flora. Combinations of phage species have been more effective than single phages. Shigella dysentery in most cases requires low doses to infect humans. Aggregates of three phages (pSf-1, pSb-1, and pSs-1) can effectively inhibit Shigella infection (Jun et al., 2016). Phages have been immobilized on electrode surfaces and have been used as bio-recognition elements to identify bacteria in electrochemical biosensing approaches (Zhou et al., 2017). Bio-conjugates for bacterial detection are made by mixing phage particles with specific nanoparticles. With this method, the detection test can be completed fast. According to Ji et al. (2021), integrating microfluidics and phage-based electrochemical sensors could be a major advancement in detecting harmful microorganisms (Richter et al., 2018).

Pathogenic bacteria in water are mostly transmitted through human rather than animal waste (Soller et al., 2010). Recent studies show that phages infect certain strains of Bacteroidetes such as *Bacteroides thetaiotaomicron* GA-17 and *B. fragilis* GB124, which can be found only in human feces. *Bacteroides fragilis* RYC2056 is highly expressed in human and animal feces. Overgrowth of filamentous bacteria is believed to be the main cause of foaming problems (Liu et al., 2015b). A phage-based technique has been proposed to address sludge swelling and foaming (Mathieu et al., 2019). *Haliscomenobacter hydrossis* is an important filamentous bacterium responsible for sludge bulking (Kotay et al., 2011). After mixing the bacteria with normal sewage sludge, phages are added to the sludge to accelerate the settling process (Kotay et al., 2011). In addition, two phages, GTE7 and GTE2 (multi-host microorganisms), represent a strong opportunity for the presence of foam (Petrovski et al., 2011a, b). Liu et al. using isolated *Gordonia* phages have tested in a simulated aeration tank system and found that the phages not only survive stably but also significantly reduce the amount of sludge sediment volume (Liu et al., 2015b). Enzymes that reduce the levels of surface polymers such as exopolysaccharides (EPS) and LPS are introduced to phages (Roach and Donovan, 2015). Transgenic phage with polysaccharide depolymerase can easily find a host receptor that is more effective in controlling biofilms (Pires et al., 2017). For example, an overexpressed T7-DspB (polysaccharide depolymerase) phage has been constructed and used to infect biofilms (Lu and Collins, 2007). The modified phage reduced the bacterial biofilm and has been more effective than the wild-type phage. Phage cocktails and multivalent phages are an important way to expand the phage host and thereby raise the efficiency of biofilm removal (Mathieu et al., 2019). In addition, nanoparticles can disrupt the biofilm and help the phage to reach relatively inaccessible sites in the biofilm, thereby improving the efficiency of the phage. Combining these new technologies with phages may be an effective way to control biofilm contamination in the near future. Phages can be formulated with certain antibiotics to improve their activity (Alisky et al., 1998; Goodridge, 2010; Kutter et al., 2010; Rastogi et al., 2018).

## 2.5 PERSPECTIVES

The future of using phages in different applications is promising. It might be a positive point that phage use has declined in the antibiotic period. Nowadays, the people are more knowledgeable. Mistakes happened due to applying the antibiotics randomly and for each infection can be avoided. Understanding lessons from our misuse to the first gift, the antibiotic will help us to maintain the second gift, the phages.

The phage applications are numerous. Meanwhile they reached different levels including the electrode and the nanoparticle applications. Genetic engineering, molecular biology, nanotechnology and the like give more chances. Meanwhile, the nature of the phages gives them their unique properties. They are viruses that could reach microscopic prokaryotic creatures. They could take the challenge and fight microbes that are invisible to us. In nature, if some parameters are adjusted, they could react positively to reduce the global waste pollution, particularly those which support the harmful bacterial growth.

On the gene level, phage genes have been cloned and used to do a specific task. For example, *E* lysis gene for bacterial host preparation shows a successful application particularly in vaccinations and drug delivery applications. In addition, not only the phage genes have been used but also the phages themselves have been engineered by adding genes to them to enable them to do additional tasks. For example, the EPS depolymerase has been cloned in the phages to be able to lyse different biofilms.

## 2.6 SUMMARY AND CONCLUSIONS

Through our optimistic venison to the future of phages, one should not neglect their negative properties such as their ability to perform horizontal gene transfer. From the different expert recommendations on antibiotic resistance and the proposed reduction strategies, reducing the phages application to a very narrow range for the most urgently needed applications might save the phage from entering the antibiotic resistance cave. In addition, using *in situ* strategy to enlarge the naturally existing phages and enabling them to control the unfavorable bacterial strains in the surrounding area will be better than seeding the environment with new phages even if they prove to be more successful. One should believe that nature can solve its problems. And one should change his own activity that accumulates new problems in the nature.

Unfortunately, the phages cannot be simply used by inexperienced persons, particularly in large applications. The microbiologists' role should not be underestimated, and they should be given the upper hand in any applications concerning the use of the phages. Meanwhile, a precise schema might be useful for nonexpert ones. From monitoring the increasing number of publications and applications concerning the phages, one could conclude that there is a solid belief that phages might be as important as antibiotics. Meanwhile the negative points that lead to the great failure in the use of the antibiotics in different applications must be avoided. To protect the future of the phages and their applications, a global comity should control their applications overall the world. Alternatively, a precise documentation containing all the required information about the phages should be available everywhere.

## REFERENCES


Abdulamir, A. S., Jassim, S. A. A., and Abu Bakar, F. (2014). Novel approach of using a cocktail of designed bacteriophages against gut pathogenic E. coli for bacterial load biocontrol. *Annals of Clinical Microbiology and Antimicrobials* **13**, 39.

Ackermann, H. W. (2006). 5500 Phages examined in the electron microscope. *Archives of Virology* **152**, 227–243.

Alexa, P., Štouračová, K., Hamřík, J., and Rychlík, I. (2001). Gene typing of the colonisation factor K88 (F4) in enterotoxigenic Escherichia coli strains isolated from diarrhoeic piglets. *Veterinární Medicína* **46**, 46–49.

Alisky, J., Iczkowski, K., Rapoport, A., and Troitsky, N. (1998). Bacteriophages show promise as antimicrobial agents. *Journal of Infection* **36**, 5–15.

Amara, A. A. A. F. (2016). Lysozymes, proteinase K, bacteriophage E lysis proteins, and some chemical compounds for microbial ghosts preparation: A review and food for thought. *SOJ Biochemistry* **2**, 1–16.

Ayyaru, S., Choi, J., and Ahn, Y.-H. (2018). Biofouling reduction in a MBR by the application of a lytic phage on a modified nanocomposite membrane. *Environmental Science: Water Research & Technology* **4**, 1624–1638.

Balogh, B., Jones, J. B., Momol, M. T., Olson, S. M., Obradovic, A., King, P., and Jackson, L. E. (2003). Improved efficacy of newly formulated bacteriophages for management of bacterial spot on tomato. *Plant Disease* **87**, 949–954.

Bao, H., Zhang, P., Zhang, H., Zhou, Y., Zhang, L., and Wang, R. (2015). Bio-control of salmonella enteritidis in foods using bacteriophages. *Viruses* **7**, 4836–4853.

Bardina, C., Spricigo, D. A., Cortés, P., and Llagostera, M. (2012). Significance of the bacteriophage treatment schedule in reducing Salmonella colonization of poultry. *Applied and Environmental Microbiology* **78**, 6600–6607.

Behle, R. W., McGuire, M. R., and Shasha, B. S. (1996). Extending the residual toxicity of bacillus thuringiensis with casein-based formulations. *Journal of Economic Entomology* **89**, 1399–1405.

Bhattacharjee, A. S., Choi, J., Motlagh, A. M., Mukherji, S. T., and Goel, R. (2015). Bacteriophage therapy for membrane biofouling in membrane bioreactors and antibiotic-resistant bacterial biofilms. *Biotechnology and Bioengineering* **112**, 1644–1654.

Brüssow, H., and Kutter, E. (2005). Phage ecology. In: *Bacteriophages: Biology and Applications*. E. Kutter and A. Sulakvelidze (eds), Boca Raton, FL: (pp 129-164) CRC Press.

Campbell, A. (1988). Phage evolution and speciation. In: *The Bacteriophages* (pp. 1–14). R. Calendar (ed). New York: Plenum Press.

Chan, B. K., and Abedon, S. T. (2012). Phage therapy pharmacology phage cocktails. In: *Advances in Applied Microbiology* (pp. 1–23). A. I. Laskin, S. Sariaslani, G. M. Gadd (eds). Amsterdam: Elsevier.

d'Herelle, F. (1917). Sur un microbe invisible antagoniste des bacilles dysent eriques. *Comptes Rendus de l'Académie des Sciences* **165**, 373–375.

Dabrowska, K., Switala-Jelen, K., Opolski, A., Weber-Dabrowska, B., and Gorski, A. (2005). Bacteriophage penetration in vertebrates. *Journal of Applied Microbiology* **98**, 7–13.

Deng, Y., Li, B., and Zhang, T. (2018). Bacteria that make a meal of sulfonamide antibiotics: Blind spots and emerging opportunities. *Environmental Science & Technology* **52**, 3854–3868.

Dias, E., Ebdon, J., and Taylor, H. (2018). The application of bacteriophages as novel indicators of viral pathogens in wastewater treatment systems. *Water Research* **129**, 172–179.

Doss, J., Culbertson, K., Hahn, D., Camacho, J., and Barekzi, N. (2017). A review of phage therapy against bacterial pathogens of aquatic and terrestrial organisms. *Viruses* **9**, 50.

Evans, T. J., Crow, M. A., Williamson, N. R., Orme, W., Thomson, N. R., Komitopoulou, E., and Salmond, G. P. C. (2010). Characterization of a broad-host-range flagellum-dependent phage that mediates high-efficiency generalized transduction in, and between, *Serratia* and *Pantoea*. *Microbiology* **156**, 240–247.

Fischetti V. A. (2008). Bacteriophage lysins as effective antibacterials. *Curr Opin Microbiol.* 11(5): 393–400.

Gabiatti, N., Yu, P., Mathieu, J., Lu, G. W., Wang, X., Zhang, H., Soares, H. M., and Alvarez, P. J. J. (2018). Bacterial endospores as phage genome carriers and protective shells. *Applied and Environmental Microbiology* **84**, e01186–18.

Gao, E. B., Gui, J.-F., and Zhang, Q.-Y. (2012). A novel cyanophage with a cyanobacterial nonbleaching protein a gene in the genome. *Journal of Virology* **86**, 236–245.

Gervais, L., Gel, M., Allain, B., Tolba, M., Brovko, L., Zourob, M., Mandeville, R., Griffiths, M., and Evoy, S. (2007). Immobilization of biotinylated bacteriophages on biosensor surfaces. *Sensors and Actuators B: Chemical* **125**, 615–621.

Goodridge, L. (2010). Designing phage therapeutics. *Current Pharmaceutical Biotechnology* **11**, 15–27.

Goodridge, L., and Abedon, S. T. (2003). Bacteriophage bcontrol and bioprocessing: Application of phage therapy to industry. *Feature Article* **53**, 254–262.

Goodridge, L., Chen, J., and Griffiths, M. (1999). Development and characterization of a fluorescent-bacteriophage assay for detection of Escherichia *coli* O157:H7. *Applied and Environmental Microbiology* **65**, 1397–1404.

Han, F., Li, J., Lu, Y., Wen, J., Zhang, Z., and Sun, Y. (2014). Isolation and characterization of a virulent bacteriophage fPA-HF17 of *Pseudomonas aeruginosa*. *International Journal Bioautomation* **18**, 241–250.

Hayashi, T. (2001). Complete genome sequence of enterohemorrhagic *Eschelichia coli* O157:H7 and genomic comparison with a laboratory strain K-12. *DNA Research* **8**, 11–22.

Henry, M., and Debarbieux, L. (2012). Tools from viruses: Bacteriophage successes and beyond. *Virology* **434**, 151–161.

Heringa, S. D., Kim, J., Jiang, X., Doyle, M. P., and Erickson, M. C. (2010). Use of a mixture of bacteriophages for biological control of Salmonella *enterica* strains in compost. *Applied and Environmental Microbiology* **76**, 5327–5332.

Hudson, J. A., Billington, C., Carey-Smith, G., and Greening, G. (2005). Bacteriophages as biocontrol agents in food. *Journal of Food Protection* **68**, 426–437.

Hudson, J. A., Billington, C., Cornelius, A. J., Wilson, T., On, S. L. W., Premaratne, A., and King, N. J. (2013). Use of a bacteriophage to inactivate Escherichia coli O157:H7 on beef. *Food Microbiology* **36**, 14–21.

Ignoffo, C. M., Garcia, C., and Saathoff, S. G. (1997). Sunlight stability and rain-fastness of formulations of baculovirus heliothis. *Environmental Entomology* **26**, 1470–1474.

Jaiswal, A., Koley, H., Ghosh, A., Palit, A., and Sarkar, B. (2013). Efficacy of cocktail phage therapy in treating Vibrio cholerae infection in rabbit model. *Microbes and Infection* **15**, 152–156.

Jassim, S. A. A., Abdulamir, A. S., and Abu Bakar, F. (2012). Novel phage-based bio-processing of pathogenic Escherichia coli and its biofilms. *World Journal of Microbiology and Biotechnology* **28**, 47–60.

Jassim, S. A. A., and Limoges, R. G. (2013). Impact of external forces on cyanophage-host interactions in aquatic ecosystems. *World Journal of Microbiology and Biotechnology* **29**, 1751–1762.

Jassim, S. A. A., and Limoges, R. G. (2014). Natural solution to antibiotic resistance: Bacteriophages 'the living drugs'. *World Journalof Microbiology* and *Biotechnology* **30**, 2153–2170.

Jassim, S. A. A., and Limoges, R. G. (2017). Bacteriophages: Practical Applications for Nature's Biocontrol. New York: Springer.

Ji, M., Liu, Z., Sun, K., Li, Z., Fan, X., and Li, Q. (2021). Bacteriophages in water pollution control: Advantages and limitations. *Frontiers of Environmental Science & Engineering* **15**, 84.

Jiang, S. C., and Paul, J. H. (1998). Significance of lysogeny in the marine environment: Studies with isolates and a model of lysogenic phage production. *Microbial Ecology* **35**, 235–243.

Jun, J. W., Giri, S. S., Kim, H. J., Yun, S. K., Chi, C., Chai, J. Y., Lee, B. C., and Park, S. C. (2016). Bacteriophage application to control the contaminated water with Shigella. *Scientific Reports* **6**, 22636.

Khan, F. M., and Gupta, R. (2020). *Escherichia coli* (*E. coli*) as an indicator of fecal contamination in groundwater: A review. In: *Sustainable Development of Water and Environment* (pp. 225–235). H-.Y. Jeon (ed). New York: Springer.

Kotay, S. M., Datta, T., Choi, J., and Goel, R. (2011). Biocontrol of biomass bulking caused by *Haliscomenobacter hydrossis* using a newly isolated lytic bacteriophage. *Water Research* **45**, 694–704.

Kutter, E., De Vos, D., Gvasalia, G., Alavidze, Z., Gogokhia, L., Kuhl, S., and Abedon, S. (2010). Phage therapy in clinical practice: Treatment of human infections. *Current Pharmaceutical Biotechnology* **11**, 69–86.

Laemmli, U. K. (1970). Cleavage of structural proteins during the assembly of the head of bacteriophage T4. *Nature* **227**, 680–685.

Larbanoix, L., Burtea, C., Ansciaux, E., Laurent, S., Mahieu, I., Vander Elst, L., and Muller, R. N. (2011). Design and evaluation of a 6-mer amyloid-beta protein derived phage display library for molecular targeting of amyloid plaques in Alzheimer's disease: Comparison with two cyclic heptapeptides derived from a randomized phage display library. *Peptides* **32**, 1232–1243.

Lin, N.-T., Chiou, P.-Y., Chang, K.-C., Chen, L.-K., and Lai, M.-J. (2010). Isolation and characterization of ϕAB2: A novel bacteriophage of Acinetobacter baumannii. *Research in Microbiology* **161**, 308–314.

Liu, H., Niu, Y. D., Meng, R., Wang, J., Li, J., Johnson, R. P., McAllister, T. A., and Stanford, K. (2015a). Control of *Escherichia coli* O157 on beef at 37, 22 and 4°C by T5-, T1-, T4-and O1-like bacteriophages. *Food Microbiology* **51**, 69–73.

Liu, M., Gill, J. J., Young, R., and Summer, E. J. (2015b). Bacteriophages of wastewater foaming-associated filamentous Gordonia reduce host levels in raw activated sludge. *Scientific Reports* **5**, 13754.

Lone, A., Anany, H., Hakeem, M., Aguis, L., Avdjian, A.-C., Bouget, M., Atashi, A., Brovko, L., Rochefort, D., and Griffiths, M. W. (2016). Development of prototypes of bioactive packaging materials based on immobilized bacteriophages for control of growth of bacterial pathogens in foods. *International Journal of Food Microbiology* **217**, 49–58.

Lu, T. K., and Collins, J. J. (2007). Dispersing biofilms with engineered enzymatic bacteriophage. *Proceedings of the National Academy of Sciences* **104**, 11197–11202.

Madhusudana Rao, B., and Lalitha, K. V. (2015). Bacteriophages for aquaculture: Are they beneficial or inimical. *Aquaculture* **437**, 146–154.

Marks, T., and Sharp, R. (2000). Bacteriophages and biotechnology: A review. *Journal of Chemical Technology & Biotechnology* **75**, 6–17.

Mathieu, J., Yu, P., Zuo, P., Da Silva, M. L. B., and Alvarez, P. J. J. (2019). Going viral: Emerging opportunities for phage-based bacterial control in water treatment and reuse. *Accounts of Chemical Research* **52**, 849–857.

Maxwell, K. L. (2016). Phages fight back: Inactivation of the CRISPR-Cas bacterial immune system by anti-CRISPR proteins. *PLoS Pathogens* **12**, e1005282.

Moye, Z., Woolston, J., and Sulakvelidze, A. (2018). Bacteriophage applications for food production and processing. *Viruses* **10**, 205.

O'Flaherty, S., Ross, R. P., Meaney, W., Fitzgerald, G. F., Elbreki, M. F., and Coffey, A. (2005). Potential of the polyvalent anti-Staphylococcus bacteriophage K for control of antibiotic-resistant taphylococci from hospitals. *Applied and Environmental Microbiology* **71**, 1836–1842.

Oliveira, M., Abadias, M., Colás-Medà, P., Usall, J., and Viñas, I. (2015). Biopreservative methods to control the growth of foodborne pathogens on fresh-cut lettuce. *International Journal of Food Microbiology* **214**, 4–11.

Pai, M., Kalantri, S., Pascopella, L., Riley, L. W., and Reingold, A. L. (2005). Bacteriophage-based assays for the rapid detection of rifampicin resistance in *Mycobacterium tuberculosis*: A meta-analysis. *Journal of Infection* **51**, 175–187.

Panthel, K., Jechlinger, W., Matis, A., Rohde, M., Szostak, M., Lubitz, W., and Haas, R. (2003). Generation of *Helicobacter pylori* ghosts by PhiX protein E-mediated inactivation and their evaluation as vaccine candidates. *Infection and Immunity* **71**, 109–116.

Petrenko, V. A., and Vodyanoy, V. J. (2003). Phage display for detection of biological threat agents. *Journal of Microbiological Methods* **53**, 253–262.

Petrovski, S., Seviour, R. J., and Tillett, D. (2011a). Characterization of the genome of the polyvalent lytic bacteriophage GTE2, which has potential for biocontrol of Gordonia-, Rhodococcus-, and *Nocardia*-stabilized foams in activated sludge plants. *Applied and Environmental Microbiology* **77**, 3923–3929.

Petrovski, S., Seviour, R. J., and Tillett, D. (2011b). Prevention of Gordonia and Nocardia stabilized foam formation by using bacteriophage GTE7. *Applied and Environmental Microbiology* **77**, 7864–7867.

Pires, D. P., Melo, L. D. R., Vilas Boas, D., Sillankorva, S., and Azeredo, J. (2017). Phage therapy as an alternative or complementary strategy to prevent and control biofilm-related infections. *Current Opinion in Microbiology* **39**, 48–56.

Pirnay, J.-P., De Vos, D., Verbeken, G., Merabishvili, M., Chanishvili, N., Vaneechoutte, M., Zizi, M., Laire, G., Lavigne, R., Huys, I., Van den Mooter, G., Buckling, A., Debarbieux, L., Pouillot, F., Azeredo, J., Kutter, E., Dublanchet, A., Górski, A., and Adamia, R. (2010). The Phage therapy paradigm: Prêt-à-porter or sur-mesure? *Pharmaceutical Research* **28**, 934–937.

Rastogi, V., Pragya, Verma, N., Mishra, A. K., Nath, G., Gaur, P. K., and Verma, A. (2018). An overview on bacteriophages: A natural nanostructured antibacterial agent. *Current Drug Delivery* **15**, 3–20.

Richter, Ł., Janczuk-Richter, M., Niedziółka-Jönsson, J., Paczesny, J., and Hołyst, R. (2018). Recent advances in bacteriophage-based methods for bacteria detection. *Drug Discovery Today* **23**, 448–455.

Roach, D. R., and Donovan, D. M. (2015). Antimicrobial bacteriophage-derived proteins and therapeutic applications. *Bacteriophage* **5**, e1062590.

Rong, R., Lin, H., Wang, J., Khan, M. N., and Li, M. (2014). Reductions of *Vibrio parahaemolyticus* in oysters after bacteriophage application during depuration. *Aquaculture* **418–419**, 171–176.

Schmelcher, M., and Loessner, M. J. (2014). Application of bacteriophages for detection of foodborne pathogens. *Bacteriophage* **4**, e28137.

Schmieger, H., and Schicklmaier, P. (1999). Transduction of multiple drug resistance of Salmonella enterica serovar typhimurium DT104. *FEMS Microbiology Letters* **170**, 251–256.

Sharaf, M. M., Salem, S. R., and Amara, A. A. (2018). Bacteriophage and bacterial lysate guarantee an additional source of proteins for the *Arthrospira fusiformis*. *Austin Journal of Microbiology* **4**, 1020.

Shkoporov, A. N., Khokhlova, E. V., Fitzgerald, C. B., Stockdale, S. R., Draper, L. A., Ross, R. P., and Hill, C. (2018). ΦCrAss001 represents the most abundant bacteriophage family in the human gut and infects Bacteroides intestinalis. *Nature Communications* **9**, 4781.

Soller, J. A., Schoen, M. E., Bartrand, T., Ravenscroft, J. E., and Ashbolt, N. J. (2010). Estimated human health risks from exposure to recreational waters impacted by human and non-human sources of faecal contamination. *Water Research* **44**, 4674–4691.

Sorokulova, I., Olsen, E., and Vodyanoy, V. (2014). Bacteriophage biosensors for antibiotic-resistant bacteria. *Expert Review of Medical Devices* **11**, 175–186.

Spricigo, D. A., Bardina, C., Cortés, P., and Llagostera, M. (2013). Use of a bacteriophage cocktail to control Salmonella in food and the food industry. *International Journal of Food Microbiology* **165**, 169–174.

Stewart, Jassim, Denyer, Newby, Linley, and Dhir (1998). The specific and sensitive detection of bacterial pathogens within 4 h using bacteriophage amplification. *Journal of Applied Microbiology* **84**, 777–783.

Sukumaran, A. T., Nannapaneni, R., Kiess, A., and Sharma, C. S. (2015). Reduction of Salmonella on chicken meat and chicken skin by combined or sequential application of lytic bacteriophage with chemical antimicrobials. *International Journal of Food Microbiology* **207**, 8–15.

Summers, W. C. (2016). Félix Hubert d'Herelle (1873–1949): History of a scientific mind. *Bacteriophage* **6**, e1270090.

Sun, W.-J., Liu, C.-F., Yu, L., Cui, F.-J., Zhou, Q., Yu, S.-L., and Sun, L. (2012). A novel bacteriophage KSL-1 of 2-Keto-gluconic acid producer Pseudomonas fluorescens K1005: Isolation, characterization and its remedial action. *BMC Microbiology* **12**, 127.

Suttle, C. A. (2000). Ecological, evolutionary, and geochemical consequences of viral infection of cyanobacteria and eukaryotic algae. In: *Viral Ecology* (pp. 247–296). C.J. Hurst (ed). Amsterdam: Elsevier.

Tan, L. T. H., Chan, K. G., and Lee, L. H. (2014). Application of bacteriophage in biocontrol of major food-borne bacterial pathogens. *Journal of Molecular Biology and Molecular Imaging* **1**, 9.

Tiwari, B. R., Kim, S., and Kim, J. (2013). A Virulent Salmonella enterica serovar Enteritidis phage SE2 with a strong bacteriolytic activity of planktonic and biofilmed cells. *Journal of Bacteriology and Virology* **43**, 186.

Twort, F. W. (1915). An investigation on the nature of ultra-microscopic viruses. *The Lancet* **186**, 1241–1243.

Verbeken, G., De Vos, D., Vaneechoutte, M., Merabishvili, M., Zizi, M., and Pirnay, J.-P. (2007). European regulatory conundrum of phage therapy. *Future Microbiology* **2**, 485–491.

Vondruskova, H., Slamova, R., Trckova, M., Zraly, Z., and Pavlik, I. (2010).  Alternatives to antibiotic growth promoters in prevention of diarrhoea in weaned piglets: A review. *Veterinární Medicína* **55**, 199–224.

Withey, S., Cartmell, E., Avery, L. M., and Stephenson, T. (2005). Bacteriophages-potential for application in wastewater treatment processes. *Science of the Total Environment* **339**, 1–18.

Wittebole, X., De Roock, S., and Opal, S. M. (2013). A historical overview of bacteriophage therapy as an alternative to antibiotics for the treatment of bacterial pathogens. *Virulence* **5**, 226–235.

Wommack, K. E., and Colwell, R. R. (2000). Virioplankton: Viruses in aquatic ecosystems. *Microbiology and Molecular Biology Reviews* **64**, 69–114.

Wong, C. L., Sieo, C. C., Tan, W. S., Abdullah, N., Hair-Bejo, M., Abu, J., and Ho, Y. W. (2014). Evaluation of a lytic bacteriophage, Φ st1, for biocontrol of Salmonella enterica serovar Typhimurium in chickens. *International Journal of Food Microbiology* **172**, 92–101.

Wu, B., Wang, R., and Fane, A. G. (2017). The roles of bacteriophages in membrane-based water and wastewater treatment processes: A review. *Water Research* **110**, 120–132.

Yang, H., Liang, L., Lin, S., and Jia, S. (2010). Isolation and Characterization of a Virulent Bacteriophage AB1 of Acinetobacter baumannii. *BMC Microbiology* **10**, 131.

Yoshida-Takashima, Y., Yoshida, M., Ogata, H., Nagasaki, K., Hiroishi, S., and Yoshida, T. (2012). Cyanophage infection in the bloom-forming cyanobacteria Microcystis aeruginosa in surface freshwater. *Microbes and Environments* **27**, 350–355.

Young, R. (1992).Bacteriophage lysis: mechanism and regulation. *Microbial Rev*, 56(3):430–491.

Zhang, A., Call, D. R., Besser, T. E., Liu, J., Jones, L., Wang, H., and Davis, M. A. (2019). β-lactam resistance genes in bacteriophage and bacterial DNA from wastewater, river water, and irrigation water in Washington State. *Water Research* **161**, 335–340.

Zhang, Z., Kottadiel, V. I., Vafabakhsh, R., Dai, L., Chemla, Y. R., Ha, T., and Rao, V. B. (2011). A promiscuous DNA packaging machine from bacteriophage T4. *PLoS Biology* **9**, e1000592.

Zhou, Y., Marar, A., Kner, P., and Ramasamy, R. P. (2017). Charge-directed immobilization of bacteriophage on nanostructured electrode for whole-cell electrochemical biosensors. *Analytical Chemistry* **89**, 5734–5741.

Zimmer, M., Vukov, N. s., Scherer, S., and Loessner, M. J. (2002). The murein hydrolase of the bacteriophage φ3626 dual lysis system is active against all tested Clostridium perfringens strains. *Applied and Environmental Microbiology* **68**, 5311–5317.

Zinno, P., Devirgiliis, C., Ercolini, D., Ongeng, D., and Mauriello, G. (2014). Bacteriophage P22 to challenge Salmonella in foods. *International Journal of Food Microbiology* **191**, 69–74.

# 3 Biodegradable Waste Management Technologies to Mitigate Climate Change

*Chaichi Devi and Meena Khwairakpam*

## 3.1 INTRODUCTION

Human activities produce waste. Waste is nothing but materials that are not useful at that moment. The same waste when managed properly may become a value-added product. The generation of waste is increasing with human advancement. Economic development leads to unavoidable by-products. The growing urban sprawl and movement is the main reason for diverse waste generation. The waste generation varies from countries to countries (Diaz-Chavez et al. 2016). But characteristics of urban waste are more or less the same for all countries, whereas they are different from rural areas. The packaging industry has contributed different elements to the waste stream. The mixture of inorganic and organic wastes is mostly seen. According to source of wastes, they can be categorized into household, municipal, biomedical, industrial, and agricultural wastes. Based on texture and basic components, a mixture of paper, plastic, cooked food, textile, glass, metal, garden waste, hazardous wastes, floral wastes, construction and demolition wastes are seen (Letcher and Vallero 2011). The mixed wastes must be segregated at source for further processing. Various rules have been implemented for segregation of wastes at source. Municipal Solid Waste (Management and Handling) Rules, 2013, came up in India and focuses on a business model. Municipalities are responsible for managing their wastes by segregating at source, door-to-door collection, sending the waste according to their characteristics for recycling, compost, and biogas facilities. The Plastic Waste (Management and Handling) Rules, 2011, has come up separately for the safe disposal of plastics. Segregation makes it easier to utilize bio-wastes in an efficient way. Recycling is emerging as green economy. The three R principles, i.e., reuse, recycle, and recover needs full enforcement and awareness. The benefits of the RRR principle are depicted in Figure 3.1. In most of the developing countries, there is lack of incentives from the policy-makers. There is requirement for the promotion of recycling business and strategies to develop training for proper management. The citizen's approach and behavior bring lots of differences. The society can make a difference by civic sense and awareness programs by municipal authorities under the government. A poster has been designed in this context by the author during a project with Pune Municipal Cooperation, India in 2010 (Figure 3.2). They are the generators and controllers at the same time. Individual awareness is enough to solve the waste management problem to a large extent. It was observed that community members are aware of pollution prevention by waste management. Strong individual commitment for waste segregation was observed in countries like Tanzania, Malaysia and Ethiopia. In China, there was strong community network influencing waste segregation (Kihila et al. 2021).

In rural areas, the biodegradable wastes are mainly agro-wastes. In certain situations, small amounts may be problematic due to lack of disposal facilities. Quantity depends on seasonal variation mainly for biodegradable wastes and the number of generators in a particular area. Broadly, waste is categorized into biodegradable and nonbiodegradable. Biodegradable waste is broken down


DOI: 10.1201/9781003386902-4

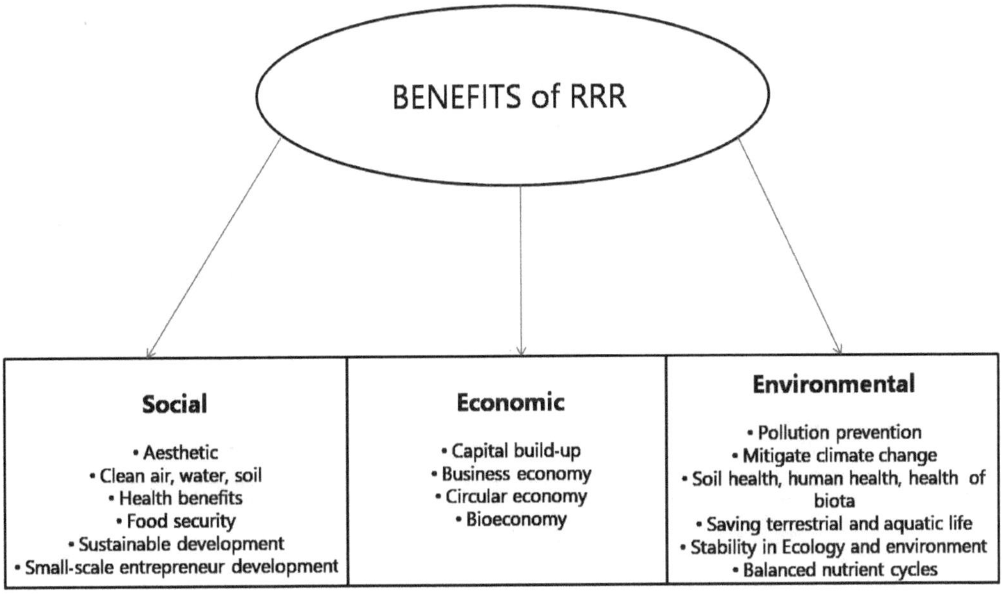


**FIGURE 3.1** Benefits of the Reuse, Recycle, Recover principle.

by microorganisms into simple molecules liberating water and carbon-di-oxide. Nonbiodegradable wastes consist of more chemicals and toxins, and cannot be decomposed. They are more harmful to the environment. The major difference is that biodegradable waste can enter the biogeochemical cycle again, whereas nonbiodegradable is always disturbing to the normal cycles (Vallero 2019). Climate is a major concern for survival, and various researches are conducted all over to combat climate change. When generally talking about carbon footprint, the first thought that comes to mind is to reduce vehicular and industrial pollution. But the contribution of waste and livestock as well toward greenhouse gas emission is large. Material management is very much important for every stage of the life-cycle of a product. Final disposal of any material discarded as wastes can become an option for recovery of value-added products or recycled into another useful product (Swati et al. 2018).

## 3.2 WASTE MANAGEMENT AND CLIMATE CHANGE: CHALLENGES

Decomposition leads to methane, carbon-di-oxide emission for bio-wastes, whereas other pollutants like NOX, halogens are released from both organic and inorganic wastes processing by various methods. The bio-wastes are not retained in the environment for long. Hence, the impact is less in comparison to other hazardous wastes. But when the global scenario is concerned, this impact locally is significant enough to alter the system. Quantifying greenhouse gas emission from waste management at different zones and stages revealed huge variations. Emission of 1,784 mg $CH_4 m^2/min$, 2.3 mg $N_2O$ $m^2/min$ and 5,448 mg $CO_2 m^2/min$ was recorded as the maximum in an open dump site (Ngwabie et al. 2019). The major issue in waste is the heterogeneous mixture leading to various obnoxious gases. The proper segregation, technology and management are the priorities. Sustainable management of wastes is the main factor to secure the planet earth and the future generation environmentally as well as economically. Waste management not only gives rise to issues related to environment but society as well. The total cost of municipal waste handling and management are around 97.57 million USD/year. The morphological diversity as well as the population regime in a

**FIGURE 3.2** A poster was designed by the author during project work with Pune Municipal Cooperation (PMC), India.

particular region impacts waste generation and transportation. A large amount of money and time is involved when there is traffic congestion and narrow roads (Barchiesi et al. 2022). Source segregation and management of biodegradable wastes is more convenient for both environment and society (Devi and Khwairakpam 2020a). Decentralized composting units and biogas plants are desirable to cut the costs in transportation and bulk waste management (Varma and Kalamdhad 2017; Saha et al. 2018; Barchiesi et al. 2022; Devi and Khwairakpam 2021).

Major inventories are required to monitor greenhouse gas emission from waste. In a recent development, South Africa has implemented inventories both at the municipal and national level. With the implementation of Clean Development Mechanism (CDM), greenhouse gas inventory was done in the landfill site. But the documentation is not appropriate due to few loopholes (Friedrich and Trois 2010). Modern inventories are required to develop an international system for better outcomes. The expertise in technology must assist in the implementation at the ground level. In another example, daily generation of wastes in Bangalore city in India is around 3,000–4,000 tons. Improved lifestyle is the main factor for increased wastes over the decades. Partial segregation with quasi centralized collection is done. But centralized system is not scientific, and there is high risk of pollution. Under "Plastic Waste Management and Handling Rules, 2011" by the Ministry of Environment and Forests, the informal recycling system is merged with the formal sector. There is mandatory coordination of the urban local body with other stakeholders including waste collectors to separate recyclable materials at transfer stations. The only challenge is the disposal of the left-over waste. Either an open dump yard or unscientific landfill is the final disposal site in all the developing countries. This has made the system uneconomical, unethical and against a healthy environment (Nedumaran and Mehala 2020). The decentralized waste management facility is easy to operate; the cost is less when managing at a small scale. In a case study from Mumbai, it is mentioned that the waste pickers are assigned to assess the worth of every waste. The waste is segregated and processed according to its end value. Mostly, the biodegradable wastes are handled in decentralized vermin composting system. Industrial waste in the Godrej society initiative for sustainable development "Good and Green" is segregated into different compartments. They hand over the wastes according to their characteristics for further processing. Around 98% wastes are recycled under the zero waste policy (Iyer 2016). From a management point of view, the burden is shared by various stakeholders. The decision-making becomes faster. The only drawback is that there is no standardized policy, and there is a chance of monopoly by raising own mini industries.

## 3.3 SOURCES OF BIODEGRADABLE WASTES AND SINK OF POLLUTANTS

In general, the picture of pollutants is that of toxins and chemicals. Few of the materials do not look like pollutants but have severe impact on the environment. The constituent material may not be hazardous but when released into the environment becomes harmful. The consequence of mismanagement, partial degradation and release of certain elements may lead to water, soil and air pollution. The integrity of the ecology and environment is disturbed due to imbalance. Everything till its need is a resource, and then it's a waste. Let us consider flowers that we use for decoration or any other purpose. A flower blooming on a plant and is plucked for decoration serves its motives. It is no longer appreciated 1 or 2 days later when it starts decaying, losing its positive characteristics and thrown away as a waste. Once it is discarded, it becomes a waste. Daily consumption pattern in a civilized society releases bulk of waste in which the constituent of biodegradable waste is very high. In India, bio-wastes comprises around 51% of the total waste. The biodegradable waste has various sources and components (Latcher and Vallero 2011). According to source, it can be categorized into household, commercial, community, agricultural, garden wastes, institutional, industrial, etc. (Barchiesi et al. 2022). The long-term impact of biodegradable wastes is not observed. The pollutants released due to improper management of wastes causes climate change gradually, but the impact on health due to contamination is immediate. As compared to the air pollution, the health impacts are mostly observed. But the impact on both environment as well as health is not negligible (Reis 2011).

Today, the generation of waste is increasing many folds. The waste generation depends on various factors like education, ethics, consumption pattern, characteristics of population, and management. But handling of waste and processing depends on the adaptation of proper technology (Swati et al. 2018). Greenhouse gases are responsible for increasing earth's temperature by preventing infrared radiation to escape outer space. Increasing greenhouse gases cause greenhouse effects. Waste contributes to net flux of greenhouse gases. The emission of greenhouse gas from wastes depends on factors like quantity of wastes, from where it is released and the type of processing of wastes. All these factors need a well-balanced policy both at the public and government level. Mcthanc is the second most abundant greenhouse gas in the atmosphere. Landfills are grounds where waste is disposed in a scientific way. Present landfills are not at all desirable in most of the developing countries. According to federal bodies, municipal landfills must be prohibited due to compacted refuse, partial degradation and poor gas trapping system (Weiner and Matthews 2003). The release of methane from landfills where mostly municipal solid wastes are dumped instead of proper disposal is a burning issue. The research is involved for a better solution. Unfortunately in developing countries like India there is no well-designed landfill site (Swati et al. 2018), whereas the amount of heterogeneous waste generation is very large as compared to developed countries. The source of waste is important to decide the management option. For example, when a small amount of waste is generated in households or commercial local units, mass burning in an open area is a common practice (Friedrich and Trois 2010). This releases pollutants which are not of much concern locally, but the net effect is huge in global climate change. The various incineration processes like gasification, mass-burning without recovery, Refuse derived fuel, and pyrolysis lead to the release of various pollutants. The energy recovery option from many of these processes is a security for the future but the pollutants released has adverse impacts. The emission from incineration plants basically includes aerosols, particulate matter, organic compounds, greenhouse gases, dioxins, furans, halogens, etc. (Reis 2011; Vallero 2011).

The sink of pollutants depends on the source and characteristics of wastes (Figure 3.3). Mostly the waste is a mixture of organic and inorganic components. The landfill is a major disposal facility for biodcgradable wastes. When the waste is not segregated before disposal, apart from methane, many other pollutants are released. The sink of methane is the atmosphere. Methane accumulated in the atmosphere is the major greenhouse gas responsible for climate change. On the other hand, when waste is incinerated, they release various air pollutants like NOX, halogens, and carbon dioxide including methane (Cossu and Stegmann 2018; Manfredi et al. 2009). All of these are absorbed into the atmosphere. Apart from these, various pollutants have their fate in the nearest water bodies disturbing aquatic life as well as percolating soil harmful to soil biota. The sink of the pollutants can be altered by little negative flux of greenhouse gases. Mostly, the technologies adopted for bio-wastes are very important. The various bio-processing facilities like bio-refining, composting, anaerobic digestion, microbial enzyme processing are probable solutions to arrest greenhouse gases. The alternative processes like composting and anaerobic digestions make the biodegradable wastes less prone to the release of greenhouse emissions (Biala 2011). The biogas produced from anaerobic digestion captures methane to be used for electricity production. On the other hand, the amount of wastes bio-processed during composting is not capable of producing methane. The stabilization of wastes makes it suitable for soil quality enhancement.

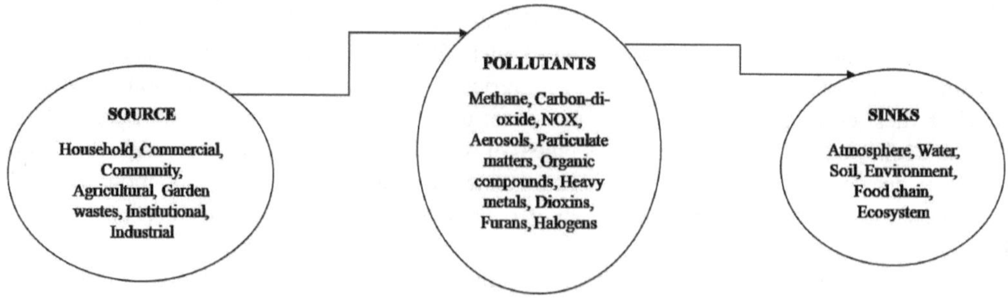


**FIGURE 3.3** Source and sink of pollutants from biodegradable wastes.

## 3.4 VARIOUS TECHNOLOGIES FOR BIODEGRADABLE WASTE MANAGEMENT

The waste generation depends on various factors like education, ethics, consumption pattern, characteristics of population, and management. But handling of waste and processing depends on the adaptation of proper technology. Biodegradation is a simple process yet the mechanism involved is complex. The various biotic and abiotic factors are involved in this process. Depending on these factors, the technologies are evolved. Few of these technologies are based on traditional practices, while modern technologies are more efficient in handling a large amount of wastes. The technologies so far used and experimented for biodegradable waste management are described in the subsequent paragraph (Table 3.1).

### 3.4.1 LANDFILL

Landfill is an engineered method for the disposal of solid wastes. In a landfill, various recovery options are possible. The only drawback is that a properly designed landfill site is rare. The present landfill in most of the developing countries is turning into a dumping ground due to lack of management. A well-fabricated landfill is the most desirable for ultimate disposal of wastes (Manfredi et al. 2009). But when the wastes are not handled and processed properly, a landfill site can be devastating to the environment. Water pollution, soil pollution as well as atmospheric pollution is mostly observed around a landfill. Pollutants in a landfill are a burning topic in current research. Various societal problems like health, aesthetic and economical problems arise due to waste disposal in a poor landfill. Especially, biodegradable wastes are not suitable for a landfill. The partial degradation of bio-wastes releases carbon-di-oxide and methane. A modern landfill covers all the aspects which are pre-treatment of the wastes, taking care of the released pollutants, impact on environment and sustainable issues. The main purpose is to give lesser burden on the environment and stabilize the wastes. The various engineering aspects like geotechnical, drainage, leachate, liner, hydraulic components must be considered. The designing of a landfill for a huge amount of wastes is a challenge (Cossu and Stegmann 2018). In a landfill where the organic load is more and basically an open dump system, the greenhouse gas emission is a major concern depending on carbon pool. Accounting up to 1,000 kg $CO_2$eq. ton (−1) in an open landfill and dispersion of methane gas are primary contributors to greenhouse gases (Manfredi et al. 2009). Landfills in developing countries are lack of base-liner, leachate and gas collection system. Toxic materials are percolating to soil and nearby water. In India, only 28% of 43 million tons of waste collected are treated well and the rest is dumped. Mostly the waste disposed is heterogeneous in nature, and bio-wastes are not treated separately. While disposing bio-wastes in a landfill, there is a high risk of various microbes and vector-borne diseases. The methane produced from incomplete biodegradation process leads to climate change. Globally, around 13% methane emission is from landfills (Swati et al. 2018).

**TABLE 3.1**
**Biodegradable Waste Management Techniques and Problems Associated**

| Technologies | Problems Associated |
|---|---|
| Landfill | Pollution, leachate, health, cost, aesthetic, social, lack of technology, poor management |
| Incineration | Pollution, health, environment, fire hazard |
| Composting and vermicomposting | Handling huge quantity of wastes, cost and space for big compost units, geographical barriers, climate, technical knowledge |
| Anaerobic digestion | Cost and space for installation, quantity of wastes, technical knowledge |
| Biorefineries | Quality of products, technological barriers, cost of production |

### 3.4.2 Incineration

Incineration is nothing but burning. Incineration is the process where combustion of wastes takes place. General idea of incineration is the waste to energy. There are various advancements in the incineration technique to produce maximum energy from a waste. The process depends on what type of waste is incinerated. Burning in open area is a common practice for small amount of wastes in households, commercial area, mainly in developing countries. But there are hazards involved related to environment, health and atmosphere in the context of incineration of wastes. The released gases have the potential to alter the environment. Bioaccumulation, long half-life period, persistence in atmosphere for longer duration brings huge impact on different ecosystem (Reis 2011). The United States have their own incineration protocol. Data must be established for a new type of waste by a burning trial before being subjected to mass incineration. Various parameters must be taken into consideration like combustion temperature, feeding rate, greenhouse gas measurement in the exhaust, operational time and equipment. If there is operational failure, there is possibly a high risk of contamination (Vallero 2011). The resource recovery is depleted in a waste mass although minimal energy is generated. The zero waste targets are difficult to achieve with these methods (Rajadesingu et al. 2021). Bio-wastes are not segregated and incinerated with other wastes in a municipal system. Mass-burning combustor in municipal facilities releases high amount of dioxin and furans. According to US Environment Protection Agency source, open burning of yard waste is the number one source of locally elevated dioxin and furan (Shields et al. 2007).

### 3.4.3 Composting and Vermicomposting

Compost is the value-added product to any organic waste. Compost is utilized as a soil enhancer. Application of compost not only provides nutrition to soil but also improves soil health and water retention capacity. Vermicomposting is a biotechnological advancement of traditional composting in which earthworms are used (Yang et al. 2014; Devi and Khwairakpam 2019). The various organic wastes are experimented for bioconversion by different composting methods. The final product is rich in nutrients, suitable for soil application. The wastes so far studied are problematic in terms of generation, handling and disposal in various geographical and ecological locations (Biruntha et al. 2020; Hanc and Dreslova 2016; Ananthavalli et al. 2019; Devi and Khwairakpam 2020a, 2021). In the recent studies, lignocellulosic biomass is of main concern as mostly agri-waste and household bio-wastes comprises lignin. Lignin is difficult to break by bioconversion (Hubbe 2014; Devi and Khwairakpam 2020b). Pre-treatment and co-composting may increase the degradation rate. Vermicompost converts lignocelluloses to lignin faster in comparison to composting. Lignin degradation is the main factor for mass loss, and mature compost retains more nutrients as compared to volatile fraction when applied to soil (Benítez et al. 2002).

### 3.4.4 Anaerobic Digestion

Traditional practice used for decades is the biggest alternative to nonrenewable energy source. Biogas is the major by-product from optimal treatment for bio-wastes. Biogas produced can be directly used or transformed into alternative energy. The efficiency of any biogas plant depends on the characteristics of the feed, temperature, and capacity (Jaysingpure and Khobragade 2023). The lignocellulosic biomass from agricultural wastes is a challenge for biogas production due to poor nutrient content and difficulty in biotransformation. A new insight in co-digestion has come up to digest two substrates together to increase efficiency. The buffering capacity can be improvised by increasing C/N ratio (Kainthola et al. 2019; Jaysingpure and Khobragade 2022). The various models are described for biomethane production by utilizing agricultural wastes. The yield depends on various factors including digestion stability, lignocellulosic mass and requirement of inoculum to increase rate (Nguyen et al. 2019).

### 3.4.5 Biorefinery

The major objective of biorefineries is to achieve decarbonization. Biorefinery is a technology to recycle biomass into either energy or products. There is limited product development from biorefineries at present. The product produced is not economical in terms of quantity and quality. The dyes and essential oils require a large quantity of raw input before being bio-pressed (Jones et al. 2022). The energy from bio-residues can increase its total value many folds. The technology can become a backbone for bio-economy. The only requirement is appropriate design and operation. Development as a rural technology can lead to sustainability (Diaz-Chavez et al. 2016). The lignocellulosic biomaterials are very suitable for the production of second generation biomethanol based on gasification technology. The other technology like biomass-to-liquid (BTL) is suitable to produce syngas from biomass (Konwar et al. 2018). Sugarcane bagasse has the potential for bioenergy generation. The prospect is large in terms of carbon storage. A large sugarcane-based biorefinery was investigated in Brazil for possible integration of carbon capture from atmosphere by bioenergy generation as a carbon sink. It was found that sugarcane bagasse bioenergy conversion was a powerful system to cut carbon emission, sustainability and economic gain (Milao et al. 2021). A business model must be developed with the integration of government policy as suggested when biorefinery prospects were analyzed in China. The new technology-based food wastes biorefinery is a green initiative in China. The advanced technologies are able to overcome the difficulties in food waste management and secure economy through lowering ecological foot prints. The financial and technical stability may bring these initiatives to the top in hierarchy to achieve sustainable development (Duan et al. 2021).

The life-cycle assessment of the raw material is very important for the selection of any of the technologies. Some of these technologies require pre-treatment depending on the bio-wastes characteristics (Jones et al. 2022). For a circular economy, it is essential to consider environmental, economic as well as societal impacts. Resources are regenerated and restored in a circular economy. The values are retained for a longer period. Reuse and redesign of a material into a new product is the basis for a circular economy. Waste is considered as a resource under sustainable material management of Environmental Protection Agency. Maximum recovery from a material is important for inclusive, equitable sustainable development and climate crisis as well (EPA 2022). Waste can be utilized to build capital. Maximum reduction of wastes and putting it back into the economy to extend its value is the essence of a circular economy. Life cycle of material continues without termination and zero waste can be achieved (Mukhopadhyay and Mukhopadhyay 2021). Circular economy is an integral part of private and public policy in many Latin American countries and The Caribbean. International trade is directly influenced by circular economy as developing countries generate most of the wastes. Waste becomes the main driver in trade. The emphasis must be given for public–private relationship for transfer of technology (Albaladejo et al. 2021). The economic flow will be smooth and circulate in the same loop without any loss of resources and money. Bioresources are managed with biotechnological processes for their economic gains. The value addition to any bio material is the biobased economy, i.e., bioeconomy. The policy Strategy on Biotechnology (2002) is based on exploring knowledge for bioeconomy. On the basis of this policy gradually Strategy on Bioeconomy came up in 2012. Sustainable bioproducts, security of job and good income were the major agendas. Biobased industries with objective for bioeconomy have emerged in Europe in around 60 projects funded to reduce dependency on fossil fuels. This will lead to sustainable development and mitigate climate change (Lago et al. 2019). In the coming decades, bio-waste will play an important role in the economy. Carbonaceous bio-waste from urban, agricultural, forest, industrial areas are being generated in large quantities globally. In the United Kingdom alone, annually over 100 million tons of municipal biodegradable wastes and 14 million tons of agricultural and forest wastes are produced. The various technologies can convert these into value-added products. Only requirement is a holistic approach, optimization of the system for maximum benefits (Guo 2018). Recycling is the major policy for EU waste target. The transition to circular economy based on bio-wastes is only possible through recycling. The target is 63% of municipal waste recycling by 2023 under waste policy in EU (Sulewski et al. 2021). The lignocellulosic materials are processed differently than the other bio-wastes (Devi and Khwairakpam 2020b). The major objective of any of these technologies is maximum recovery and economic gain. The most challenging is the implementation with regard to geographical location, government protocol, social objection, law and all the stakeholders (Diaz-Chavez et al. 2016). The landfill option is more desirable when greenhouse gases are concerned. But fabrication of a high-quality landfill site is a big challenge. Soil quality improvement, increasing carbon storage on top soil, recovery of resources including climate change mitigation only can be achieved through alternative options to landfill. Both from economic and sustainable points of view, the options of composting and anaerobic digestions are considered the best (Saha et al. 2018).

## 3.5 VALUE-ADDED PRODUCTS

Obtaining value-added products is the primary consideration of any waste processing. The potential of biodegradable waste to generate various value-added products makes it a resource for sustainable development. It can be recycled, reused or recovered. The technology depends on the nature of the substrate as well as decomposition time. Various researches are conducted to find the most efficient technology for a specific waste to be processed. The initial characteristic of the waste is the major criteria for experimentation (Devi and Khwairakpam 2019). The management cost to handle a large amount of wastes includes its collection, transportation and disposal. In the transit, few options may be beneficial as the waste is actually a nation's wealth. The transfer stations are there where the waste is segregated for its recycle and recovery option. The biodegradable wastes are best suited for recovery of various valuable products (Biala 2011). The waste is then transferred to management option depending on their characteristics. The most significant is the processing of bio-wastes at its source when possible. The bulk amount of waste is important when we want to recover for big economic gain. The most resilient sustainable approach is the processing of biodegradable wastes for a circular bio-economy. The various value-added products like biogas, compost, vermicompost, recycled items, energy and products from biorefineries are obtained from biodegradable wastes (Usmani et al. 2021; Devi and Khwairakpam 2020b; Kainthola et al. 2019; Konwar et al. 2018). Mostly the biogas production in large quantity is always desirable. On the other hand, compost and vermicompost are always preferable for marginal farmers and small-scale industries. The energy recovery from agriculture based bio-wastes is more in comparison to other fuel-based materials (Saha et al. 2018). It can be utilized as the second-generation bi-fuels. All these methods are eco-friendly and economical. These small-scale industries can benefit the marginal society to a large extent. Research on the exploitation of agro-wastes to value-added products is still very limited. The green circular economy can be achieved from this sector (Babu et al. 2022).

## 3.6 SUSTAINABLE FUTURE FROM WASTE MANAGEMENT

The three R options, i.e., reuse, recycle, recovery are all applicable to biodegradable wastes. Recycling into value-added products has maximum environment and economic gain. Carbon emission is the major challenge for climate change. Low carbon emission for green economy with various strategies is important for sustainable development. Underlying mechanisms must be explored to tackle a big problem. If a large amount of organic waste is recycled into a value-added product, there will be significant reduction in methane emission. Organic farming is the best option to sequester carbon. Carbon and nutrient cycling is maintained through soil amendment with compost (Devi and Khwairakpam 2021). Sequestering carbon through compost is an important mitigation measure for climate both in the long and medium terms. The percentage retention of carbon through compost in soil is able to reduce greenhouse gas to the maximum within 20 years. Compost application is beneficial to carbon credit and clean air development. As per reports from Germany, the

United Kingdom, Australia and North America, the carbon is increased by 60 (36–84) kg/ha/yr per tons of compost applied. On the other hand, curbing greenhouse gas is possible by cutting transportation and manufacture of chemical fertilizer by on-site decentralized compost facility (Biala 2011).

The bioenergy production is a noble option for clean energy. The study of five African economies responsible for 60% carbon emission has forwarded measures to reduce carbon emission by 2030. Carbon abatement came up through policy of source and production of biofuels (Okorie and Lin 2022). Biogas produced from anaerobic digestion of various biomass is an accepted alternative fuel source where nonrenewable resources are exhausted (Saha et al. 2018; Kainthola et al. 2019). The sustainability issue to secure future generation depends on alternative source of energy and food security. Both of these objectives to meet the sustainable development goals can be achieved through alternative management options of biodegradable wastes. The biomass generation is a concern for its sustainable utilization. The bio-economy concept is fully realized when a waste is subjected to life-cycle assessment and techno-economic analysis to optimize bio-processing for maximum value (Usmani et al. 2021).

## 3.7 BIODEGRADABLE WASTES AND THEIR MANAGEMENT

The generation of biodegradable waste has various sources as mentioned earlier. The experimentation is going on to find various alternative options. Wide varieties of organic wastes are subjected to decomposition for the production of various value-added products. The results are impressive for practical implications. The degradation process depends on the climatic conditions and geographical locations. In the household sector, vegetable wastes, cooked food and other materials are segregated from the waste stream for resource recovery. In the industrial sector, paper and pulp industry, sugar cane industry, dairy, textiles and many others generates bio-wastes. These are experimented in various time and space for their management. Agricultural area produces various by-products. Agricultural residues, weed biomass, and leaf litter are common biodegradable wastes recovered in fields. The potential of lignocellulosic biomass to generate bioenergy and biofertilizers is large. Proper management can recover a good return from all these wastes. Few of the technologies adopted for various bio-wastes are discussed below and summarized in Table 3.2.

Industrial wastes are banned from dumping in landfill in various countries. London convention held in Korea imposed a ban on discarding paper mill waste in any landfill site (Lee et al. 2017). The waste generated in such industries poses a threat to the environment. Food industry, paper mill industries, textile industries, sugar industries and many others generate bio-wastes. Bakery industry

**TABLE 3.2**
**Technologies Studied for Few Bio-wastes**

| Bio-wastes | Technology |
|---|---|
| Paper mill wastes | Composting, anaerobic digestion |
| Sugarcane bagasse | Composting, anaerobic digestion |
| *Hydrilla verticillata* aquatic weed biomass | Composting |
| Mikania micrantha weed | Composting |
| Lantana camara weed | Vermicomposting |
| Parthenium hysterophorus weed biomass | Composting, vermicomposting, anaerobic digestion |
| Ageratum conyzides weed biomass | Vermicomposting, anaerobic digestion |
| Saccharum spontaenum weed biomass | Vermicomposting |
| Seaweeds | Composting, vermicomposting |
| Agricultural residues | Composting, vermicomposting, anaerobic digestion |
| Municipal bio-wastes | Composting, vermicomposting, anaerobic digestion |

waste was subjected to vermicompost. The biotransformation helped in nutrient recovery and minimized load on the disposal site. The industries may set up their own compost plants within the premises for decentralized management (Yadav and Garg 2019). In the pulp and paper industry, the organic residual is generated from wastewater treatment. Various studies have shown that composting is the most acceptable solution for paper mill wastes. Pile composting, rotary drum composting, windrow composting even in pilot scale is successful. Various amendments are done with paper mill sludge to achieve optimum nutrient recovery. The final C/N ratio depends on the initial mixing ratio. The compost thus produced is stable. The paper mill sludge compost from windrow was compared with commercial fertilizers. The nutrient supplement was more in compost than commercial supplements. It is capable for the growth of various horticultural plants during pot experiments (Evanylo and Daniels 2013). The physico-chemical analysis proves that the particle size of the final compost and its porosity is ideal for plant growth. The composted sludge possesses less phytotoxicity than the raw material (Lee et al. 2017). Pilot scale studies have shown the potential for anaerobic digestion. The microbial community is adapted to remove toxicity by co-digestion. The focus must be on operational dynamics of bioengineering for further research on lignocellulosic biomass (Meyer and Edwards 2015). Sugarcane baggage is another raw material for composting. It can be completely broken down and is ideal for commercial composting unit. The addition of sugarcane baggage in mixed green wastes increases coarseness and reduces particle size, balances pH, and increases enzymatic activities and nutrient values. Co-composting always benefits the substrate biodegradation. The pre-composting reduces time (Melati et al. 2017). The combination of windrow and vessel composting, vessel and vermicomposting has scope to manage large quantity of wastes (Varma and Kalamdhad 2017). Anaerobic digestion is another way to utilize sugarcane bagasse for biogas production. Biomethane production is enhanced by various pre-treatment and co-digestion methods. The maximum recovery is possible through advanced technologies like carbonization, pyrolysis and gasification with anaerobic digestion. The general practice for sugarcane bagasse is burning, but with the valorization process, sustainable management can be achieved (Agarwal et al. 2022). The biofuels like ethanol and xylitol are obtained by enzymatic hydrolysis after pre-treatment. The global energy demand can be met through sugarcane bagasse utilization (Ajala et al. 2021). Municipal solid waste was extensively studied in various environments for value addition. The growing population and consumption pattern have contributed a lot to the waste stream in urban areas (Kaviraj 2003). Cities are dependent on nearby rural areas for most of its resources. Similarly, the organic fraction of municipal waste can become a resource for the nearby agricultural lands if properly managed. The recovery option of compost and vermicompost is more due to high organic carbon around 18%–36%. Vermicompost was found to be more effective for plant growth in comparison to compost with lower C/N ratio (Tognetti et al. 2005; Ducasse et al. 2022). The major problem of municipal solid waste is its voluminous nature. There is no need for transporting biodegradable waste to a long distance once it is segregated. The decentralized facility for bioconversion can reduce the volume. It was observed that there was 60% volume reduction under optimum environmental conditions within 32–34 days. The total organic carbon was reduced, thereby preventing greenhouse gas emissions from partial degradation in landfill site (Chaudhari et al. 2011). Weeds are invasive plant species spreading in a wide area due to morphological and physiological adaptations. Many adversities are linked with invasive species including climate change. Theses invasive species have the capacity to alter the ecosystem (Simberloff 2001). The native species are replaced by the invasion in various geographical locations. Aquatic as well as terrestrial weeds are found according to their habitats. The management traits are inadequate in various ecological regimes (Batish et al. 2004; Kurien and Ramasamy 2006; Saha et al. 2018; Jain and Kalamdhad 2019; Devi and Khwairakpam 2019, 2021). Utilization of diverse weed biomass for the production of compost, vermicompost and biogas was extensively studied in various regions (Yadav and Garg 2011; Singh and Kalamdhad 2016; Jain and Kalamdhad 2019; Ananthavalli et al. 2019; Kauser et al. 2020; Devi and Khwairakpam 2021; Saha et al. 2022). Weeds are problems in urban areas, agriculture land, and forest (Kohli et al. 2006). In India, three most abundant weeds, namely, *Ageratum conyzides*, *Lantana camara*, and *Parthenium hysterophorus*, in agriculture, forest, and urban land, respectively, were processed through vermicomposting experimentation (Devi and Khwairakpam 2020a, b, 2021). Vessel rotary drum composting of *Mikania micrantha* weed biomass took 20 days to complete degradation. C/N ratio decreased up to seven and nutrient values increased. Phytotoxicity assessment on different plants gave positive results (Kauser et al. 2020). Rotary drum composter was favored to manage aquatic weed *Hydrilla*. Wetlands are biologically diverse and sensitive ecosystems. *Hydrilla* is abundant in many wetlands across the countries disturbing aquatic life. It is very difficult to eradicate and to dispose the weed biomass. The stabilized compost was obtained after 20 days of rotary drum composting. The physicochemical properties were suitable for field applications. This can be a novel approach for *Hydrilla* management (Jain and Kalamdhad 2018). The management and handling depends on the physics of compost. The compost reduces the volume of wastes and reduces pressure on transportation as well. This is crucial for any land applications. Vermicompost application was tried in various plant growths for its nutritional property. The plant growth and nutrient recovery was beyond marginal (Ghadge and Jadhav 2013; Hanc and Chadimova 2014; Sharma et al. 2016). The compost was analyzed for physical properties. There was around 64% reduction in bulk density and increase in free air space. The organic amendments make the compost better as a soil enhancer (Jain and Kalamdhad 2019). Similarly, vermicomposting of various weeds species have shown nutritional enhancement in the final product. The value-added vermicompost prepared from weed biomass is ideal for further soil applications. Vermicomposting is not as rapid as the composting process but the quality of vermicompost is better than compost (Fornes et al. 2012; Devi and Khwairakpam 2020a). The combination of composting and vermicomposting technologies makes the time shorter for the complete degradation. Vermicomposting followed by composting was successful for many of the substrates (Varma and Kalamdhad 2017; Kauser and Khwairakpam 2022).

## 3.8 SUMMARY AND CONCLUSIONS

Biodegradable waste must be considered as resources for societal, economic and environmental benefits. These are the backbones of sustainability. Zero waste is difficult to achieve without recycle or recovery option from wastes. The complete circular economy depends on a product being utilized till its grave. Putting back into the cycle is only possible for biodegradable wastes. The well balanced nutrient cycle is possible from the application of organic fertilizer. This is an important aspect to maintain the ecosystem for sustainable development. The incineration and landfill depletes the recovery option. On the other hand, composting, anaerobic digestion, and biorefinery processes add value to the wastes. Waste management is an important aspect for climate change mitigation. Instead of zero waste if we target zero emission then decentralized management options are always preferable. Roadmap to a green future depends on the practical applicability to utilize waste as a resource. The initiatives from the entire stakeholders with firm policies are the only solution to prevent any adverse impact from waste and its utilization as a resource.

## REFERENCES


Agarwal, N.K., Kumar, M., Ghosh, P. 2022. Anaerobic digestion of sugarcane bagasse for biogas production and digestate valorization. *Chemosphere*. 295: 133893.

Ajala, E.O., Ighalo, J.O., Ajala, M.A. et al. 2021. Sugarcane bagasse: A biomass sufficiently applied for improving global energy, environment and economic sustainability. *Bioresources and Bioprocessing*. 8: 87. doi:10.1186/s40643-021-00440-z.

Albaladejo, M., Mulder, N., Olmos, X. et al. 2021. International trade and the circular economy in Latin America and the Caribbean. In: *Inclusive and Sustainable Industrial Development Working Paper Series* 3/2021. ECLAC-UNIDO: Vienna.

Ananthavalli, R., Ramadas, V., Paul, J.A.J. et al. 2019. Seaweeds as bioresources for vermicompost production using the earthworm, *Perionyx excavatus* (Perrier). *Bioresource Technology*. 275: 394–401.

Babu, S., Rathore, S.S., Singh, R. et al. 2022. Exploring agricultural waste biomass for energy, food and feed production and pollution mitigation: A review. *Bioresource Technology*. 360: 127566. doi:10.1002/9781119679523.ch11

Barchiesi, M.A.; Costa, R.; Di Pillo, F. 2022. The link between the compliance with environmental legislation on separate collection and the municipal solid waste costs. *Sustainability*. 14: 5661. doi:10.3390/su14095661.

Batish D.R., Singh, H.P., Kohli, R.K et al. 2004. Management of invasive exotic weeds requires community participation. *Weed Technology*. 18: 1445–1448

Benítez, E., Sainz, H., Melgar, R. et al. 2002. Vermicomposting of a lignocellulosic waste from olive oil industry: A pilot scale study. *Waste Management & Research*. 20(2): 134–142. doi:10.1177/0734242X0202000205.

Biala, J. 2011. *The Benefits of Using Compost for Mitigating Climate Change*. NSW Office of Environment and Heritage Report OEH 2011/0385. doi:10.13140/RG.2.1.1547.1126.

Biruntha, M., Karmegam, N., Archana, J. et al. 2020. Vermiconversion of biowastes with low-to-high C/N ratio into value added vermicompost. *Bioresource Technology*. 297: 122398.

Chaudhari, R.D., Datar, M.T., Babookani, M.R. 2011. Municipal solid waste management through vermicomposting employing exotic species of earthworm Eudrilus eugeniae. *Journal of Environmental Science and Engineering*. 53(1): 129–134.

Cossu, R., Stegmann, R. 2018. *Solid Waste Landfilling: Concepts, Processes, Technology*. Elsevier, Amsterdam.

Devi, C., Khwairakpam, M. 2019. Bioconversion of Lantana camara by vermicomposting with two different earthworm species in monoculture. *Bioresource Technology*. 296: 122308.

Devi, C., Khwairakpam, M. 2020a. Feasibility of vermicomposting for the management of terrestrial weed *Ageratum conyzoides* using earthworm species *Eisenia fetida*. *Environmental Technology & Innovation*. 18: 100696.

Devi, C., Khwairakpam, M. 2020b. Management of lignocellulosic green waste *Saccharum spontaenum* through vermicomposting with cow dung. *Waste Management*. 113: 88–95.

Devi, C., Khwairakpam, M. 2021. Management of invasive weed *Parthenium hysterophorus* through vermicomposting using a polyculture of *Eisenia fetida* and *Eudrilus eugeniae*. *Environmental Science and Pollution Research*. 28: 29710–29719. doi:10.1007/s11356-021-12720-4.

Diaz-Chavez, R., Stichnothe, H., Johnson, K. 2016. Sustainability considerations for the future bioeconomy. In: Lamers P., Searcy E., Richard Hess J., Stichnothe H. (eds), *Developing the Global Bioeconomy*. Academic Press, London, pp. 69–90.

Duan, Y., Chen, H., Liu, T. et al. 2021. Food waste biorefinery: Case study in China for enhancing the emerging bioeconomy. In: Bhaskar P., Varjani S., Pandey A., Rene E.R. (eds), *Waste Biorefinery*. Elsevier, Amsterdam, pp. 421–438. doi:10.1016/B978-0-12-821879-2.00015-6.

Ducasse, V., Capowiez, Y., Peigné, J. 2022. Vermicomposting of municipal solid waste as a possible lever for the development of sustainable agriculture. A review. *Agronomy for Sustainable Development*. **42**: 89. doi:10.1007/s13593-022-00819-y

EPA. 2022. *EPA Circular Economy Conference*. https://www.epa.gov/recyclingstrategy/what-circular-economy.

Evanylo, G.K., Daniels, W.L. 2013. Paper mill sludge composting and compost utilization. *Compost Science & Utilization*. 7(2): 30–39.

Fornes, F., Mendoza-Hernandez, D., Garcia-de-la-Fuente et al. 2012. Composting versus vermicomposting: A comparative study of organic matter evolution through straight and combined processes. *Bioresource Technology*. 118: 296–305.

Friedrich, E., Trois, C. 2010. Greenhouse gases accounting and reporting for waste management: A South African perspective. *Waste Management*. 30(11): 2347–2353. doi:10.1016/j.wasman.2010.05.016.

Ghadge, S., Jadhav, B., 2013. Effect of Lantana manures on nutrient content of Fenugreek (*Trigonella-foenum graecum* L.). *Bioscience Discovery*. 4(2): 189–193.

Guo, M. 2018. Multi-scale system modelling under circular bioeconomy. *Computer Aided Chemical Engineering*. 43: 833–838.

Hanc, A., Chadimova, Z., 2014. Nutrient recovery from apple pomace waste by vermicomposting technology. *Bioresource Technology*. 168: 240–244.

Hanc, A., Dreslova, M. 2016. Effect of composting and vermicomposting on properties of particle size fractions. *Bioresource Technology*. 217: 186–189.

Hubbe, M. 2014. Lignocellulose biodegradation in composting. In: Maheshwari D. (ed), *Composting for Sustainable Agriculture*. Springer, Cham, pp. 43–66. doi:10.1007/978-3-319-08004-8_3

Iyer, H. 2016. Case study of mumbai: Decentralised solid waste management. *Procedia Environmental Sciences*. 35: 101–109.

Jain, M.S., Kalamdhad, A.S. 2018. Efficacy of batch mode rotary drum composter for management of aquatic weed (*Hydrilla verticillata* (L.f.) Royle). *Journal of Environmental Management*. 221: 20–27. doi:10.1016/j.jenvman.2018.05.055.

Jain, M.S., Kalamdhad, A.S. 2019. Drum composting of nitrogen-rich *Hydrilla Verticillata* with carbon-rich agents: Effects on composting physics and kinetics. *Journal of Environmental Management*. 231(3): 770–779.

Jaysingpure, V.D., Khobragade, M.U. 2023. Biogas generation through anaerobic digestion of organic waste: A review. In: Ranadive M.S., Das B.B., Mehta Y.A., Gupta R. (eds), *Recent Trends in Construction Technology and Management*. Lecture Notes in Civil Engineering. Springer, Singapore, p. 260. doi:10.1007/978-981-19-2145-2_49.

Jones, R.E. Speight, J.L., Blinco, I.M. et al. 2022. Biorefining within food loss and waste frameworks: A review. *Renewable and Sustainable Energy Reviews*. 154: 111781.

Kainthola, J., Kalamdhad, A.S., Goud, V.V. 2019. A review on enhanced biogas production from anaerobic digestion of lignocellulosic biomass by different enhancement techniques. *Process Biochemistry*. 84: 81–90.

Kauser, H., Khwairakpam, M. 2022. Organic waste management by two-stage composting process to decrease the time required for vermicomposting. *Environmental Technology and Innovation*. 25: 102193.

Kauser, H., Pal, S., Haq, I. et al. 2020. Evaluation of rotary drum composting for the management of invasive weed Mikania micrantha Kunth and its toxicity assessment. *Bioresource Technology*. 13: 123678.

Kaviraj, S.S. 2003. Municipal Solid Waste management through vermicomposting employing exotic and local species of earthworms. *Bioresource Technology*. 90: 169–173.

Kihila, J., Wernsted, K., Kaseva, M. 2021. Waste segregation and potential for recycling: A case study in Dar es Salaam City, Tanzania. *Sustainable Environment*. 7: 1. doi:10.1080/27658511.2021.1935532.

Kohli, R.K., Batish, D.R., Singh, H.P. et al. 2006. Status, invasiveness and environmental threats of three tropical American invasive weeds (Parthenium hysterophorus L., Ageratum conyzoides L., Lantana camara L. *India Biological Invasions*. 8: 1501–1510.

Konwar, L.J., Mikkola, J., Bordoloi, N. et al. 2018. Sidestreams from bioenergy and biorefinery complexes as a resource for circular bioeconomy. In: Bhaskar T., Pandey A., Venkata Mohan S., Lee D., Khanal S.K. (eds). *Waste Biorefinery*. Elsevier, Amsterdam, pp. 85–125

Kurien, J., Ramasamy, E.V., 2006. Vermicomposting of Taro (*Colocasia esculenta*) with two epigeic earthworm species. *Bioresource Technology*. 97(11): 1324–1328.

Lago, C., Caldés, N., Lechón, Y. 2019. Sustainability of bioenergy. In: *The Role of Bioenergy in the Bioeconomy*. Academic Press, London, pp. 225–296. doi:10.1016/B978-0-12-813056-8.00006-6.

Lee, J., Kim, C., Kown S. et al. 2017. Effect of composting of paper mill sludge for land spreading. *Nordic Pulp and Paper Research Journal*. 32(4): 691–698.

Letcher, T., Vallero, D. 2011. *Waste*. Elsevier, Amsterdam.

Manfredi, S., Tonini, D., Christensen, T.H. et al. 2009. Landfilling of waste: Accounting of greenhouse gases and global warming contributions. *Waste Management and Research*. 27(8): 825–836. doi:10.1177/0734242X09348529.

Melati, R.B., Alison, A.S., Fernando, C.P. et al. 2017. Sugarcane bagasse: Production, composition, properties, and feedstock potential. In: Murphy R. (ed.), *Sugarcane: Production Systems, Uses and Economic Importance*. Nova Science Publishers, Inc., Hauppauge, NY

Meyer, T., Edwards, E.A. 2015. Anaerobic digestion of pulp and paper mill wastewater and sludge. *Water Research*. 65: 321–349.

Milao, R.F.D., Araujo, O.Q.F., Medeiros, J.L. 2021. Sugarcane-based ethanol biorefineries with bioenergy production from bagasse: Thermodynamic, economic, and emissions assessments. In: Bhaskar T., Varjani S,, Pandey A., Rene, E.R. (eds), *Waste Biorefinery*. Elsevier, Amsterdam, pp. 125–158. doi:10.1016/B978-0-12-821879-2.00005-3.

Mukhopadhyay, B.R., Mukhopadhyay, B.K. 2021. What is the circular economy? *The Sentinel, Editorial*.

Nedumaran, G., Mehala, D. 2020. Waste management initiatives in India. *International Journal of Scientific & Technology Research*. 9(3): 145–149.

Nguyen, D.D., Jeon, B., Jeung, J.H. et al. 2019. Thermophilic anaerobic digestion of model organic wastes: Evaluation of biomethane production and multiple kinetic models analysis. *Bioresource Technology*. 280: 269–276.

Ngwabie, N.M., Wirlen, Y.L., Yinda, G.S. et al. 2019. Quantifying greenhouse gas emissions from municipal solid waste dumpsites in Cameroon. *Waste Management*. 15(87): 947–953. doi:10.1016/j.wasman.2018.02.048.

Okorie, D.I., Lin, B. 2022. Africa's biofuel energy and emissions prospect: Forward-looking into 2030. *Sustainable Energy Technologies and Assessments*. 53(D): 102775.

Rajadesingu, S., Deepankara, V., Dowlath, M.J.H. et al. 2021. Modern society and zero waste tools. In: Hussain C.M. (eds), *Concepts of Advanced Zero Waste Tools*. Elsevier, Amsterdam, pp. 181–213.

Reis, M.F. 2011. Solid waste incinerators: Health impacts. In: Nriagu J.O. (eds), *Encyclopedia of Environmental Health*. Elsevier, Amsterdam.

Saha, B., Devi, C., Khwairakpam, M. et al. 2018. Vermicomposting and anaerobic digestion: Viable alternative options for terrestrial weed management: A review. *Biotechnology Reports*. 17: 70–76.

Saha, B., Koley, S., Khwairakpam, M., Kalamdhad, A.S. 2022. Comparative study between mono-digestion and co-digestion of terrestrial weed (*Parthenium hysterophorus*). *Cleaner Engineering and Technology*. 11: 100560.

Sharma, R., Dwivedi, H.S., Dwivedi, P. 2016. Utilization of three obnoxious weeds (Parthenium *hysterophorus*, *Lantana camara* and *Eichhornia crassipes*) through vermicomposting and their response on vegetative growth of soybean crop. *International Journal of Advanced Research in Biological Sciences*. 3(9): 13–20.

Shields, W.J., Tondeur, Y., Benton, L., Edwards, M.R. 2007. Dioxins and furans. In: Morrison R.D., Murphy B.L (eds), *Environmental Forensics*. Academic Press, London, pp. 293–312.

Simberloff, D. 2001. Biological invasions-how are they affecting us, and what can we do about them? *Western North American Naturalist*. 61 (3):308–315.

Singh, W.R., Kalamdhad, A.S. 2016. Transformation of nutrients and heavy metals during vermicomposting of the invasive green weed *Salvania natans* using *Eisenia fetida*. *International Journal of Recycling of Organic Waste in Agriculture*. 5: 205–220.

Sulewski, P., Kais, K., Goła´s, M. et al. 2021. Home bio-waste composting for the circular economy. *Energies*. 14: 6164. doi:10.3390/en14196164.

Swati, Thakur, I.S., Vijay, V.K. et al. 2018. Scenario of landfilling in India: Problems, challenges, and recommendations. In: *Handbook of Environmental Materials Management*. pp. 1–16. doi:10.1007/978-3-319-58538-3_167-1.

Tognetti, C., Laos, F., Mazzarino, M.J. et al. 2005. Composting vs. vermicomposting: A comparison of end product quality. *Compost Science and Utilization*. 13: 6–13.

Usmani, Z., Sharma, M., Awasthi, A.K. et al. 2021. Bioprocessing of waste biomass for sustainable product development and minimizing environmental impact. *Bioresource Technology*. 322: 124548.

Vallero, D.A. 2011. Thermal waste treatment. In: Letcher T.M., Vallero D.A. (eds), Waste. Academic Press, London, pp. 219–231.

Vallero, D.A. 2019. Waste and Biogeochemical Cycling, 2nd edn. Academic Press, London, pp. 91–125.

Varma, S., Kalamdhad, A.S. 2017. *Environmental Pollutants and their Bioremediation Approaches*, 1st edn. CRC Press, London. doi:10.1201/b22171.

Weiner, R.F., Matthews, R.A. 2003. Solid waste disposal. In: *Environmental Engineering*, 4th edn. Butterworth-Heinemann, Oxford, pp. 263–271.

Yadav, A., Garg, V.K., 2011. Vermicomposting: An effective tool for the management of invasive weed *Parthenium hysterophorus*. *Bioresource Technology*. 102: 5891–5895.

Yadav, A., Garg, V.K., 2019. Biotransformation of bakery industry sludge into valuable product using vermicomposting. *Bioresource Technology*. 274: 512–517.

Yang, J., Lv, B., Zhang, J. et al. 2014. Insight into the roles of earthworm in vermicomposting of sewage sludge by determining the water-extracts through chemical and spectroscopic methods. *Bioresource Technology*. 154: 94–100.

# 4 Rehabilitation of Ecosystem Services in Hard Climate Change

*Hayfa Rajhi and Anouar Bardi*

## 4.1 INTRODUCTION

The ecosystem losses, damage and damage manifested by global climate change observed recently are the result of complex and interdependent physical and socio-economic processes, which have been at work for many decades, if not centuries. This anthropogenic climate change determines the nature and intensity of the climate-related hazards that humanity faces, together with the geographical location of the planet. The risk also depends on the exposure of human and natural systems to the hazard in question. Finally, it depends on the country's vulnerability to various hazards.

The interactions between these elements, which act on interconnected systems, can trigger a cascade of major risks in the different sectors and regions. Storm surges, floods and sea level rise can disrupt livelihoods. Systemic risks from extreme weather events can also lead to the collapse of infrastructure networks and essential services; to risks of food and water insecurity and the pandemic (the most recent one of coronavirus disease); and to the disappearance of rural livelihoods and incomes, in particular for the poorest populations. In this context, a rehabilitation of ecosystems that consists in restoring the functioning of the ecosystem of a damaged or degraded environment must be discussed. In this study, a synthesis of study cases has been presented and discussed. These cases presented clear and immediate solutions to confront the global climate change.

## 4.2 GLOBAL CLIMATE CHANGE: MANIFESTATION AND SPIN OFF

### 4.2.1 Demonstration

Loss and damage are the result of complex and interrelated physical and socio-economic processes that have been at work for many decades, if not centuries.

The buildup of greenhouse gases (GHGs) in the atmosphere will increase warming and cause lasting change across the planet. The Intergovernmental Panel on Climate Change (IPCC) is the most authoritative source for projections of climate-related hazards induced by climate change. It predicts with certainty that the impacts of climate change will increase in severity, frequency and magnitude as global warming continues, and that these impacts may become irreversible (Pachauri and Meyer, 2014). Therefore, extreme weather events may occur. Extreme weather events include increased frequency and severity of heat waves, droughts, cyclones, extreme precipitation, extreme sea levels, floods (caused by extreme precipitation, extreme sea levels and melting glaciers) and vegetation fires (resulting from a combination of factors such as heat, lack of rain and wind), among other examples. Extreme weather phenomena climate change leads to changes in the frequency, intensity, spatial extent, duration and temporal occurrence of extreme weather events, potentially leading to unprecedented extremes (UNDRR, 2015). Changes have been observed in many extreme

DOI: 10.1201/9781003386902-5

weather events since about the middle of the 20th century. Each further increase in global warming leads to a clearly noticeable increase in the intensity and frequency of heat extremes, including heat waves, heavy rainfall and sea heat waves. It also leads to an increase in the proportion of intense tropical cyclones (Pachauri and Meyer, 2014).

As shown in Figure 4.1, slow onset phenomena include sea level rise, rising temperatures, ocean acidification, retreating glaciers and related effects, salinization, land and forest degradation, biodiversity loss and desertification (UNDRR, 2015).

1. *Rising temperatures*: The temperature on the surface of the globe increased by 1.09°C in 2011–2020 compared to 1850–1900, but not all regions are experiencing the same warming. Significantly greater temperature increases on land (1.59°C on average) than on the oceans (0.88°C on average). The Polar Regions are also experiencing greater warming than the tropics, with the rise in Arctic temperatures more than twice the global average warming. Changes due to rising temperatures include heat waves and changes in ecosystem functioning (particularly in high latitudes) (GIEC, 2021).
2. *Sea level rise*: Current levels of anthropogenic sea level rise are primarily the result of thermal expansion of seawater caused by rising temperatures, melting glaciers and ice sheets making a growing contribution to the phenomenon. During the 20th century, sea level rise reached 1–2 mm/year in most regions, but the rate accelerated to 3.7 mm/year between 2006 and 2018 (GIEC, 2021).

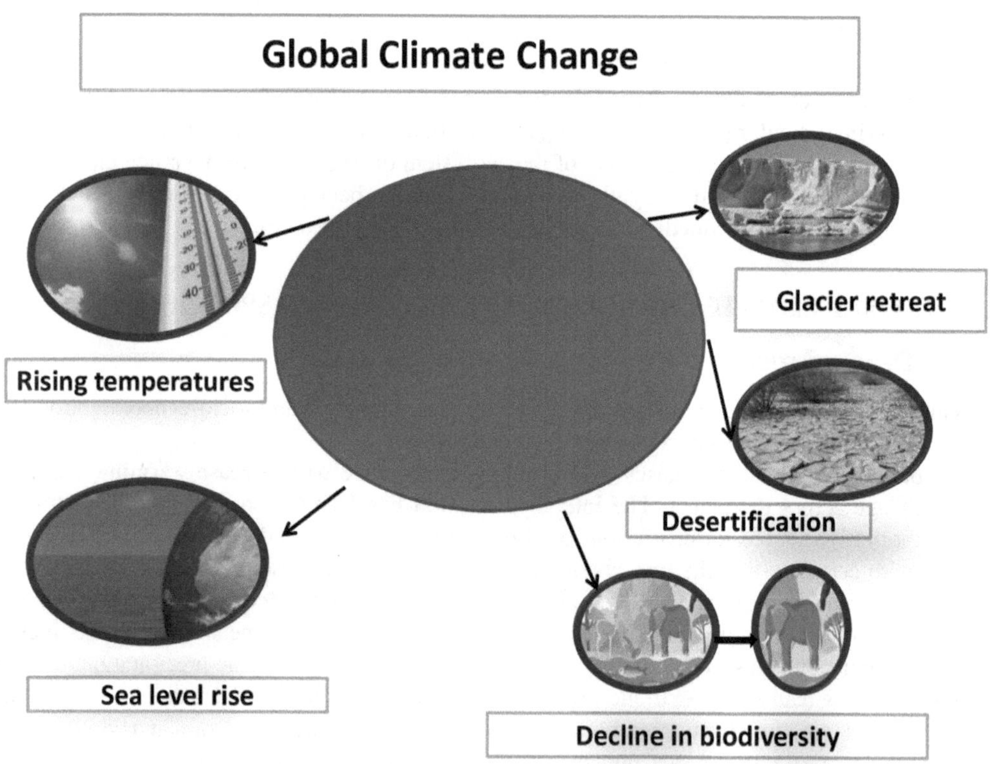


**FIGURE 4.1** Global climate change: manifestation and spinoff.

3. *Salinization*: In the case of salinization, nonsaline soil becomes sufficiently saline to inhibit plant growth, a process mainly due to sea level rise and irrigation. The main consequences of salinization are land degradation and desertification, decline in biodiversity and adverse effects on agricultural production, freshwater resources and health. It is estimated that salt affects 7.4% of the land in the world (GIEC, 2021).
4. *Ocean acidification*: Carbon dioxide ($CO_2$) in the atmosphere forms a weak acid when it dissolves in seawater. As a result, the pH of the ocean decreases as concentration of atmospheric $CO_2$ increases, which is detrimental to marine life. One of the notable consequences of ocean acidification is coral bleaching. Over the past three decades, the pH of seawater has decreased by 0.017–0.027 per decade due to increasing $CO_2$ concentrations in the atmosphere (GIEC, 2021; Zhu and Liu, 2020).
5. *Glacier retreat*: Glaciers retreat when the mass of snow and ice from the glaciers melts at a rate faster than that of their accumulation. The result is a change in the flow of meltwater rivers, which has negative effects on the availability of water for irrigation and contributes to sea level rise. In particular, the vast ice sheets of Greenland and Antarctica and the high mountain areas of the Andes, Himalayas and Alps contribute about 1.81 mm to sea level rise each year. The retreat of glaciers can have local and regional impacts on river flows, ecosystems and agricultural livelihoods (Zhu and Liu, 2020; GIEC, 2021).
6. *Land and forest degradation*: Land degradation refers to a negative trend in land holdings and the condition of land, often manifested by a reduction or loss of biological productivity, ecological integrity and/or of value to humans. Land degradation affects some 3.2 billion people worldwide. Land and forest degradation can have a wide range of impacts on the natural environment and society (e.g. loss of ecosystem services) (Pecl et al., 2017).
7. *Desertification*: This is the degradation of land in arid, semi-arid and sub-humid-dry areas, which results from the interaction of various human and environmental processes, in particular drought. Its main impacts are related to the loss of ecosystem services and its consequences for the livelihoods of populations dependent on natural resources (Pecl et al., 2017).
8. *Decline in biodiversity*: Biodiversity is the variability of living organisms in terrestrial, marine and other aquatic ecosystems. It includes variability at the genetic level, at the species level and at the ecosystem level. Biodiversity declines when variability at any of these levels decreases. The decline in biodiversity can lead to a loss of ecosystem functions. This leads to a decline in ecosystem services, such as carbon sequestration and the ability to adapt to further climate change. The main drivers of biodiversity loss are land-use change, overexploitation of animals and plants (including their illegal trade), pollution, invasive nonnative species and increasingly, the land change (Pecl et al., 2017; Guillod et al., 2017).

### 4.2.2 General Ecosystem Impacts

Soil degradation is a process that results from human productive activity and/or climate change that lowers the current and/or future capacity of the soil to support other activities necessary for human life. In general, in a hot climate, and especially when agricultural residues are removed and animal manure does not return to the soil, the organic matter content drops below 0.5%, thus deteriorating the soil structure and their fertility. In addition, rainwater clogs the surface of the soil, infiltration decreases, runoff and erosion starts, and then accelerates. In fact, soil degradation is generally a complex phenomenon, in which several elements can intervene and contribute to the loss of agricultural potential: erosion and removal of soil by water or wind, loss of fertility resulting in chemical, physical and biological changes (Desertification Report, 2015).

#### 4.2.2.1 Desertification

Soil is a living medium, and it takes about 500 years to form 5 cm of soil. More than 3.5 billion hectares of land have been degraded worldwide and may no longer be suitable for cultivation in the future. In the dry lands that occupy more than 44% of the Earth's emerged surface, the phenomenon is more marked and massive. In Africa, for example, it affects 65% of arable land (Desertification Report, 2015). This process is called "desertification". The United Nations Convention to Combat Desertification defines this term as "land degradation in arid, semi-arid and dry sub-humid areas, resulting from various factors including climatic variations and human activities". Desertification diminishes every second the productive potential of soils at all latitudes. The causes of desertification are multiple, with human activities being the biggest causal factor: excessive clearing, excessive plowing, exposure of soil fauna to air and sun, excessive irrigation, spreading of pesticides and chemical fertilizers, destruction of fauna, monoculture, loss of soil humus and loss of fertility (Desertification Report, 2015). According to the IPCC, the soil is considered to be the second carbon sink after the oceans, which can contribute massively to the mitigation and adaptation to climate change. Similarly, the United Nations Committee on Food Security (CFS) focuses on the links between soil health and the improvement of ecological and human health. Thus, and given that nearly half of the cultivated systems on earth are located in arid areas, food security is threatened by this contained degradation of agricultural soils. In addition, the UN estimates that by the year 2050, desertification, water shortages and the decline in agricultural production could lead to the forced migration of 50–700 million people (Desertification Report, 2015).

#### 4.2.2.2 Effect on Agriculture (Case of the Olive Tree): The Soil Situation in North Africa

In Tunisia, drought and desertification remarkably affect arid and semi-arid areas characterized by unfavorable climatological and hydrological conditions (URAM, 2010). The annual amount of rainfall is generally low, with a very irregular pattern in autumn and winter and only very rarely exceeding 200 mm (Floret and Pontanier, 1982). The arid regions of southern Tunisia are naturally very sensitive to wind erosion. Not only is the rainfall in these regions low, but the soils are fine, sandy and shallow, i.e., easily erodible by the wind. Similarly, groundwater also undergoes these fluctuations, which leads to the scarcity of water resources in these regions. The Tunisian south is characterized by sandy to sandy-loamy soils dedicated to fruit growing, in particular olive trees, fig trees and almond trees. Sandy soil is defined as soil that contains 85% or more sand, with a silt content of more than 1.5 times, while clay not exceeding 15% (Balba, 1995). Many problems can be raised that can be attached to sandy soils: their coarse texture is responsible for the high speed of infiltration, these soils contain little organic matter and fine particles, their ability to cation exchange is low, and the bulk density of sandy soils is generally high, being 1.6–1.8 g/cm$^3$ (Mahmoud, 2011). Finally, these soils hydrate faster but hold water less firmly than finer textured soils. Because of this, they give up water to the crop more readily than fine-textured soils. Thus, plants growing on these sandy soils generally suffer from drought unless they are frequently irrigated (Mahmoud, 2011). According to the IPCC, the soil is considered to be the second carbon sink after the oceans, which can contribute massively to the mitigation and adaptation to climate change (Desertification Report, 2015). Similarly, the United Nations CFS focuses on the links between soil health and the improvement of ecological and human health. Thus, and given that nearly half of the cultivated systems on earth are located in arid areas, food security is threatened by this contained degradation of agricultural soils. In addition, the UN estimates that by the year 2050, desertification, water shortages and the decline in agricultural production could lead to the forced migration of 50–700 million people (Desertification Report, 2015).

Soil degradation is a major obstacle to any form of economic development in the regions affected by this phenomenon. Consequently, this scourge is at the origin of the reduction or even the destruction of the biological potential of the earth, and can ultimately lead to the general degradation of ecosystems (Mahmoud, 2011). In addition, and under more severe climatic conditions, the risks for the olive sector will increase, especially with scarce rainfall, which may compromise its economic viability (Comino et al., 2021).

## 4.3 SOLUTION WITHIN THE FRAMEWORK OF A SUSTAINABLE ENVIRONMENT

### 4.3.1 Valorization of Various Wastes for a Green and Blue Economy

#### 4.3.1.1 Green Economy

*Effects of Olive By-product Compost on the Physical, Chemical and Biological Properties of Amended Soil*

The olive tree is grown on a large scale in semi-arid areas under a Mediterranean climate, characterized by long periods of soil water deficit which are observed more particularly in the period from May to October (Mezghani et al., 2012). According to the studies of Ben Rouina et al. (2002) and Ben Rouina (2007) on the vegetative cycle of the olive tree, four phases can be identified: two phases of vegetative growth and two phases of rest of the plant. The vegetative growth phases are, in turn, divided into two types: (i) the phase of intense partial vegetative growth and (ii) the phase of partial vegetative growth. The resting phases of the plant take place in two distinct periods. The first resting phase that happens in the summer period is known as the summer resting phase and the second occurs in the winter. The phase of intense vegetative growth and the partial phase will take place in spring and autumn, under favorable climatic conditions. Thus, it is necessarily the favorable environmental conditions that determine the stage of growth of the olive tree. During the remaining phases, the olive tree must adapt to harsher climatic conditions in order to overcome these detrimental survival mechanisms characterized by high temperature occurring in summer and/or low temperature (cold) occurring in the winter period. Likewise, the examination of the biological cycle of the olive tree shows that its strong activity is concentrated during the phases of vegetative and reproductive growth which occur in certain parts simultaneously. This coincidence could lead to the establishment of competition between the different tissues developed during the phenological stages (Ben Rouina et al., 2002; Ben Rouina, 2007).

The importance of olive production in terms of quantity and quality depends closely on the fertility of the soil, the climatic regime and the cultivation techniques applied. The absence of organic inputs and the practice of surface plowing frequently in the olive groves of the arid regions of Tunisia have favored the depletion and structural degradation of the soil. This has consequently increased their susceptibility to erosion by wind and water. The use of soil amendment with vegetable water and/or compost from olive tree by-products constitutes an alternative remedy to restore the status of humus in these lands, especially since the availability of manure has become increasingly rare and insufficient. The effect of using raw vegetable water as an organic soil amendment has been studied. In addition, a study related to the effect of using compost based on olive by-products as an organic amendment on the fundamental properties of the soil was discussed. The orientation toward the manufacture of compost based on a mixture of by-products of the olive tree with other organic products available in the region is of multiple interests since it allows to preserve the environment and to have an amendment product analogous to farmyard manure. In addition, this operation contributes effectively to carbon sequestration in the soil and to mitigating greenhouse gas emissions responsible for climate change.

The main objective of the work carried out is to study the effects of a compost made from the mixture of olive by-products with various other organic wastes on the essential properties of a light soil representative of the soil cover of olive groves. The results obtained showed that the addition of the manufactured compost significantly improves the physical and chemical properties of the amended soil compared to the control soil (unamended soil). It is clear that the composting of olive

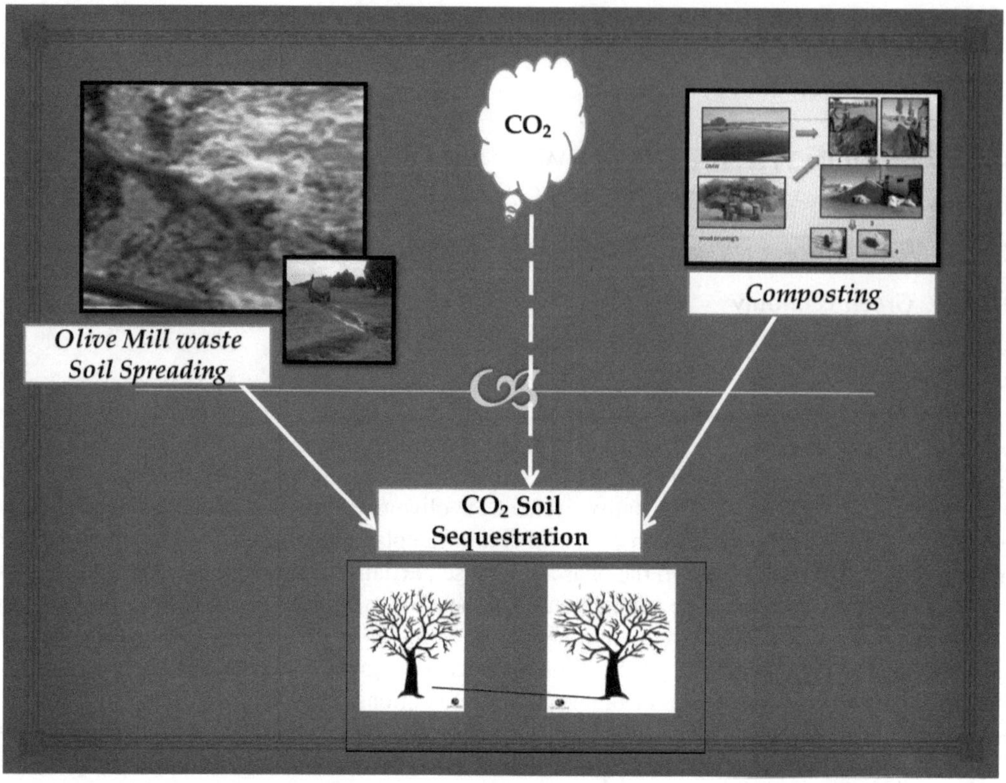


**FIGURE 4.2** Valorization of various wastes in soil composting efficiency.

by-products mixed with other organic products available in the study region offers multiple ecosystem services in a context of climate change causing more and more scarcity of natural resources (water, soil, biodiversity). Indeed, composting offers an organic product of guaranteed effectiveness in improving the fertility and productivity of soils cultivated in arid environments. Also, compost improves the structural stability of the soil and increases its resistance to degradation by water and wind erosion and by compaction. This would allow the preservation of the soil and the environment (Figure 4.2). The use of manufactured compost and its incorporation into the soil constitute a carbon sequestration procedure and an ecosystem service of great importance in the fight against atmospheric warming. Composting olive tree by-products mixed with seaweed and fish by-products also has the advantage of preserving the natural environment from pollution by these products. At the scale of olive-growing farms, this composting process constitutes a very interesting and promising alternative for the valorization of by-products of the olive tree as an amendment product for the improvement and maintenance of soil fertility and ensures sustained production. This would avoid or minimize the use of chemical fertilizers, which are expensive for farmers. In fact, the price of a ton of compost made by farmers in the study area is estimated at around \$64/ton against a price of around \$213/ton of chemical fertilizers. However, the acceptability by farmers of this type of amendment remains limited at the moment. A scientific article has been submitted in an impacted international journal.

### *Use of Olive By-products in the Manufacture of Silage for Livestock Feed in Arid Areas of Tunisia*

The olive sector generates several quantities of by-products, which have harmful effects on the environment. The by-products of the olive tree, namely, olive pomace, vegetable waters and pruning wood (James, 1990) represent 90% of the biomass offered by the olive tree. However, the recovery of these by-products remained limited in quantitative terms, and they were only subject to traditional operating methods. However, it is possible to use these by-products as raw material for obtaining new, more advantageous products such as the manufacture of silage for animal feed. These by-products contain energy, protein and fiber, which could be useful for ruminants. However, the direct use of these by-products for animal feed was limited due to low digestibility and high content of polyphenols, which decrease protein availability and microbial protein synthesis. Indeed, the ensilage of olive by-products by anaerobic fermentation (silage) has a beneficial effect for the refining and processing of these by-products by making them more effective for feeding ruminants. This procedure provides a social and environmental benefit, particularly important in semi-arid and arid ecosystems where available pasture and forage are limited. It provides an additional source of food for livestock on the one hand and avoids environmental pollution on the other.

Attempts to valorize this biomass of the olive tree, representing both economic and environmental benefits, have been initiated for several years by many organizations, and have given convincing results. In this context, our orientation has turned toward the treatment and use of different olive by-products, namely, vegetable water, pomace, pruning wood, and olive leaves as a source of food for the 'breeding. Pomace was used for its high water and lipid content. Vegetable waters have been used for their high protein and nutrient content. Pruning timbers were used because of their high lignin content. Olive leaves have been used for their interesting antibacterial, antifungal and anti-inflammatory powers (Figure 4.3).

At the economic level, several experimental works have been carried out to valorize these by-products and evaluate the socio-economic and environmental benefits of this ensilage process. In this study, we will propose to evaluate the interest of using the by-products of the olive tree as fodder resources and to enhance their socio-economic benefits, by comparing the cost of production of animal feed by the sector silage and through the barley sector. The calculation of the various

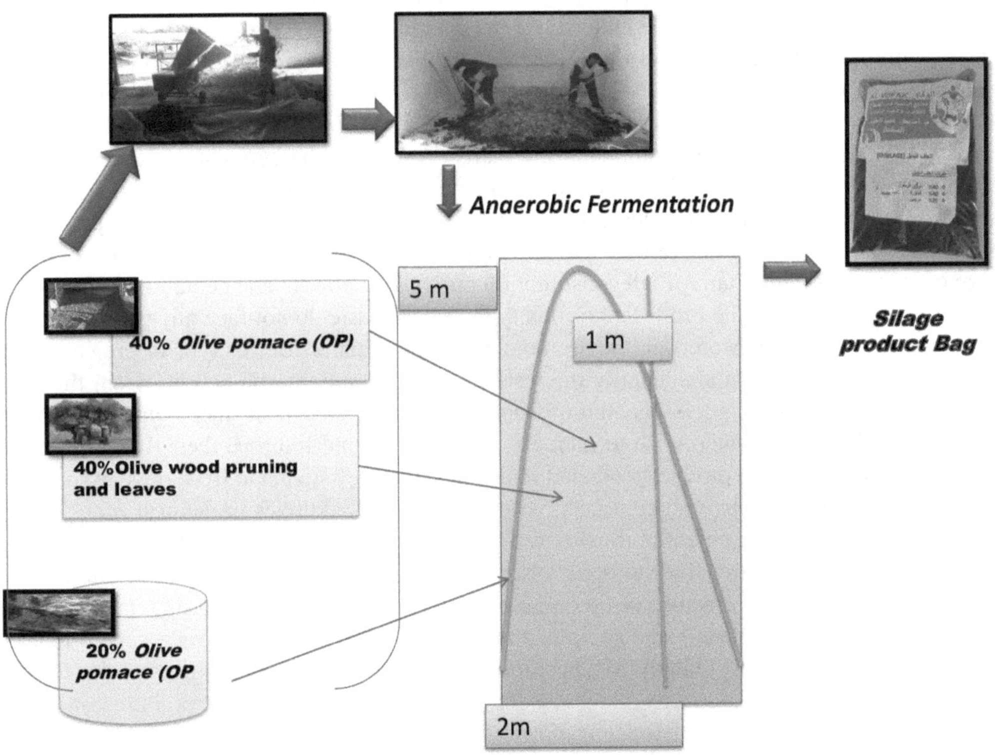


**FIGURE 4.3** Valorization of agriculture waste in silage product.

costs was carried out in US dollars (USD). Thus, compared to barley, which has a nutritional value and a fairly high selling price on the market (around 152 USD per ton of product), the selling price of a ton of silage offered on the market (as a substitute for a ton of barley) is around 25 USD. From a nutritional point of view, it has been decided that 2 tons of silage are needed to have the equivalent of the intake of 1 ton of barley. In terms of expenses, it takes about 50 USD for the 2 tons of silage. It costs 1/3 of the cost for the purchase of a ton of barley. However, it is possible to further reduce the cost of making silage. On this subject, today there is a technique of mechanized pruning (with a chainsaw) which lends itself well in areas where the olive tree is old. This technique reduces the cost of pruning, which can go from 1.8 USD per tree for manual intervention to 0.73 USD per tree for mechanized intervention. This wet silage is an efficient and economical technique which makes it possible to store very variable quantities from a few tons to several hundred tons. With this silage technique, it was possible to create a certified label for livestock, such as "fattened with natural silage". Despite these multiple advantages, the use of olive leaves and twigs in animal feed is up against several difficulties, including the dispersion of the leaves and twigs on the plots, the need to transport these by-products which increases the final cost, the existence of a pathogenic risk and problem of conservation of the leaves and risk of degradation of the food value (Rajhi et al., 2022).

#### *Other Uses of Olive By-products: Use of Olive Tree By-products and Green Clay in the Manufacture of Organic Toothpaste*

A valuation of the olive leaves of the variety of southern Tunisia (*Olea europaea* L) associated with other natural products, namely, green clay and cloves, has been the subject of production of organic toothpaste. The manufacturing protocol is made from the following elements: Olive leaves thanks to their antibacterial, antifungal and anti-inflammatory properties. Clay (green clay, sieved and purified from the Douirate-Southern Tunisian region) is used as a support to make the toothpaste. The comparison between the characteristics of the manufactured toothpaste and the international toothpaste standards shows that the majority of the characteristics of the manufactured toothpaste comply with the standards. As a result, the absence of foaming power is very respectful to health, especially it can be a remedy for the problem of dental hydration, especially of the gums caused by foaming power. Similarly, the lack of foaming power of this pharma-cosmetic product can represent a very advantageous environmental alternative to reduce the cumulative effect caused by the increase in releases of pharma-cosmetics of chemical origin on our planet. This manufactured toothpaste can enter the list of green economy products. The main objective of this project was to produce toothpastes from olive leaves as the main additive meeting the expectations of a wide range of customers with a fairly high consumption purchasing power. The demand for this type of product is quite high and the producers of this type of toothpaste do not face any problems due to its large-scale marketing. Moreover, the promoters like the initiators of this type of project may be women artisans who are fundamentally interested in starting a domestic activity from their own region. In fact, interesting socio-economic benefits can be derived from these types of projects, especially for housewives who wish to increase their income and improve their living conditions and social well-being. The product produced represents a toothpaste made from olive by-products, namely, olive leaves and olive oil combined with other natural additives in the arid zone, namely, green clay. The production potential illustrates the low cost of manufacturing the toothpaste.

This low cost therefore enables to offer a more competitive price. Several merits can be demonstrated through the manufacture of this type of toothpaste, namely, the merits related to health impacts, environmental impacts and socio-economic impacts. In fact, toothpaste has no foaming power: toothpaste can be a remedy for the problem of dental hydration. In reality, the foaming properties of toothpastes are strictly dangerous for people with disorders of the oral mucosa (canker sores, for example). The lack of foaming power of this cosmetic pharmaceutical product corresponds to a very advantageous environmental alternative to overcome the cumulative effect caused by the increase in the release of cosmetic and chemical substances into the natural environment.

The manufacture of this type of product fits perfectly with the recent evolution of the economics of sustainable development in which the environmental costs of growth are taken into account and where promoters advocate green growth, as well as the sustainability of economic development by preserving the surroundings, safeguarding the interests of future generations and improving the welfare of individuals. Green business can be defined as all economic activities that are possible to produce goods and services that help to avoid, reduce or eliminate harmful effects on the environment. Areas such as the management of scarce resources, renewable energies, climate change, risk prevention or even waste management are found in this economy. This toothpaste has been produced exclusively from natural products (Figure 4.4). This allows the exploitation of the biomass of the olive tree, formerly discarded in nature; the preservation of the environment; and the reduction of environmental costs related to the production of olives. It is remarkable that the production of olives brings (apart from its natural extracts such as olive pomace and vegetable waters) natural by-products resulting from the pruning and uprooting of trees as to their renewal by replanting. They correspond to the leaves, twigs and large wood that are generally discarded in the wild and essentially constitute a significant risk of environmental degradation. These extracts which are natural and inexpensive represent a rich source of interest for the manufacture of pharmaceutical and cosmetic products. It also represents the subject of a constructive valorization since the valorization of this biomass of the olive tree has many advantages both at the economic and environmental level. It should also be noted that the quantities available vary considerably from year to year and that the values provided are only approximate. Annually, 5 kg/year/tree of leaves, prune wood, and twigs is obtained from fall and sapling. These advantages can be summarized by the following points: (i) control of affordable energy linked to relatively low production costs compared to other types of toothpaste. Indeed, various organic toothpastes, which have been generated from very rare medicinal plants, require intensive cultivation. Both natural products have been extracted from the natural study area, which represents cheap and abundant natural resources. (ii) The manufacture of this product does not require a very long manufacturing time compared to other methods of

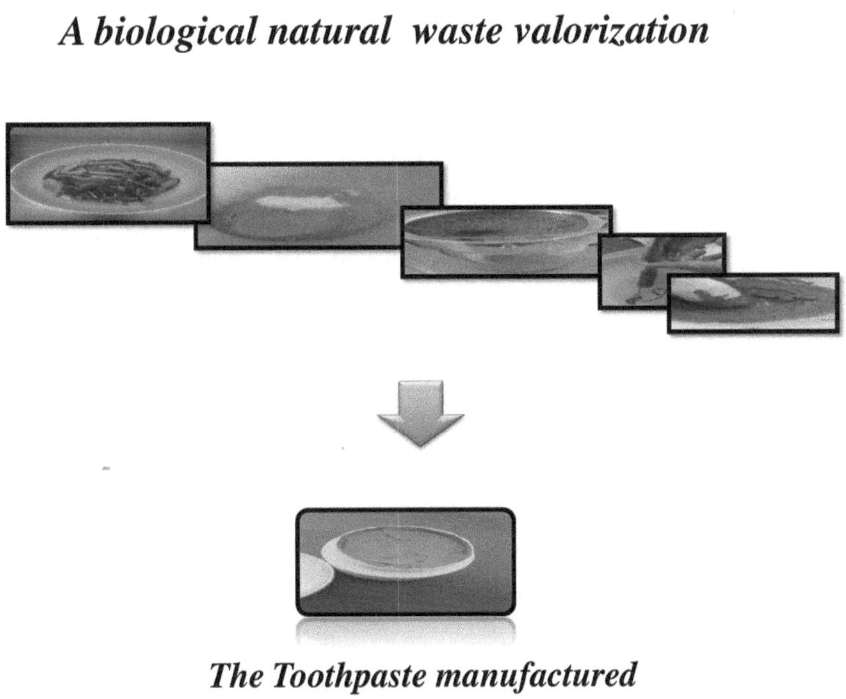


**FIGURE 4.4** Valorization of natural abundant waste in biological toothpaste.

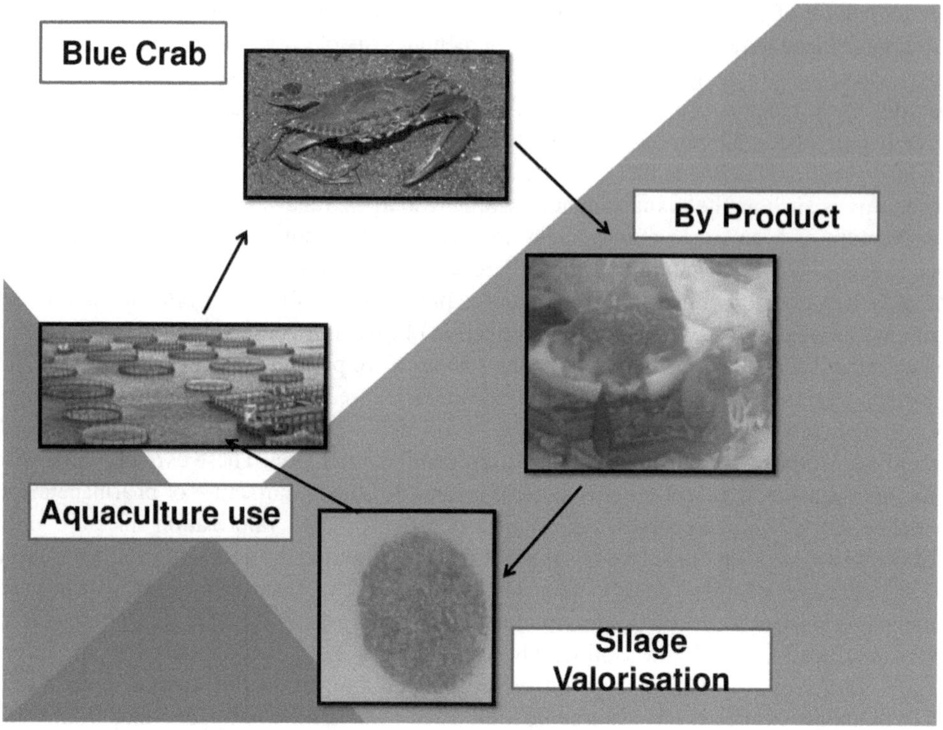


FIGURE 4.5 Recovery of crab by-products in aquaculture silage.

manufacturing dentifrices based on chemical reagents. Indeed, the chemical stability of chemical-based toothpastes requires a very long manufacturing time. Therefore, a considerable advantage in terms of profitability of economy of scale has been added to other economic advantages associated with this new product. (iii) The creation of a promising new sector for qualified entrepreneurs. (iv) A positive effect on the region, in terms of job creation compared to other types of traditional productive activities. (v) Likewise, the bearers and initiators of this type of project can be craftswomen belonging to the same region who are interested in starting a domestic toothpaste manufacturing activity. This reflects the exceptional socio-economic interest of this type of project, in particular for housewives who aspire to increase their income and improve their living conditions as well as social well-being (throughout the year) (Rajhi et al., 2021) (Figure 4.5).

#### 4.3.1.2 Blue Economy: Recovery of Crab By-products (Crab Waste)

The blue crab, an invasive species, has been widely recorded in different Mediterranean regions. Over the past decade, ecological and economic impacts have been little studied for this crustacean, native to the western shores of the Atlantic Ocean. More specifically, this species is characterized by high fecundity and aggressive behavior. However, the blue crab has rapidly developed important new supports in recent years of fishing in the Mediterranean, including the Tunisian south coast (Gulf of Gabes). In addition, blue crab represents an opportunity of economic interest by producing a valuable compound with wide applications in the food, pharmaceutical, biomedical, cosmetic, agricultural and biotechnological fields. Crabs have the distinction of being higher in various minerals, and are also rich in water, protein-free nitrogen, tryptophan, phosphorus and vitamins. In this regard, and in order to maintain the socio-economic interest of the blue crab, it is necessary to assess and control the biological and chemical contamination of the crab. There are different ways of recovering co-products. The recovery of co-products leads to derived products that are used as raw materials or ingredients in other fields such as agriculture, food and feed, pharmaceuticals, cosmetics and medicine. From the same type of co-product, it is possible to obtain different by-products: flour, oil, hydrolysates, etc. (Dumay, 2006). Studies have shown that meals from fish co-products are as effective as those from whole fish for farming shrimp cod (Tibbetts et al., 2006) and drum red (Li et al., 2004). However, fish industry processing (case of blue crab) generates a huge amount of waste, creating disposal and landfill problems. In this sense, we can develop a line of research consisting in valuing marine by-products; in this case, crab shells after their crushing and their uses as breeding livestock for small pelagics (meal for fish) and/or as microalgae nutrients, given the importance of the role of these microorganisms in the phytoremediation of heavy metals. Similarly, the use of crab waste (shell and/or gonads) can be considered, given their high levels of phosphorus and nitrogen in composting. Similarly, it becomes all the more interesting if we manage to isolate bacteria and identify them in order to valorize them later in bio-preservation processes. These bacteria can be extracted and isolated from crab by-products. As a result, marine by-products can be used in various fields ranging from returning to the ground to extracting molecules of interest for the pharmaceutical industry. Crab by-products can be used in various fields ranging from the return as breeding livestock of small pelagics (meal for fish) and/or as nutrients from microalgae to the extraction of molecules of interest for the industry.

## 4.4 SUMMARY AND CONCLUSIONS

We believe that the research work carried out is an important contribution in the direction of research on the valorization of agriculture waste, as well as the optimization of processes on all links in the value chain, while highlighting international standards of respect for the environment which aim to preserve the environment and reduce the environmental costs linked to production in the olive sector.

However, further research is worthwhile to identify various other aspects: (i) A trial over an extended period testing the impact of different doses of vegetable water and compost as a single amendment and as a combined amendment on soil fertility is essential. (ii) The study of the effect of the biochar amendment (obtained from agriculture waste) on the properties of the soil and on the production of the olive tree is also essential. (iii) The impact of the introduction of silage (made from olive by-products) in livestock feed on milk and meat production also deserves to be studied. A detailed study of the economic and environmental profitability of the different uses considered in this study is of great importance to better convince and encourage operators and decision-makers.

## REFERENCES

Balba, A.M. (1995). Management *of Problem Soils in Arid Ecosystems*. Lewis Publishers, Egypt.

Ben Rouina, B., Boukhris, M., Trigui, A. (2002). Effect of the climate and the soil conditions on crops performance of the Chemlali de Sfax olive trees. *Acta Horticulturae* 586(1), 285–289.

Ben Rouina, B. (2007). Etude des relations sol-eau-plante et de l'effet du stress hydrique sur la croissance et la fructification de l'olivier (Oleaeuropaea L.). Ph.D. thesis, Faculté des Sciences de Sfax, Sfax.

Comino, J.R., Senciales-González, J.M., Yu, Y., Salvati, L., Giménez-Morera, A., Cerdà, A. (2021). Long-term changes in rainfed olive production, rainfall and farmer's income in Bailén (Jaén, Spain). *Euro-Mediterranean Journal for EnvironmentalIntegration* 6, 58.

Desertification Report. (2015). *International Forum: Land Degradation and Climate Change, 10, 11,12,* 13, June 2015. Desertification Report, Montepellier.

Dumay, J., 2006. *Extraction de lipides par bioréacteur enzymatique combiné à l'ultrafiltration: Application à la valorisation de co-produits de poisson (Sardina pilchardus) Ecole doctorale Mécanique, Thermique et Génie civil de Nantes.* Université de Nantes, Nantes.

Floret, Ch., Pontanier, R. (1982). Aridity *in Pre-Saharan Tunisia, Climate, Soil, Vegetation and Development Work Document*. OROSTOM, Paris.

GIEC. (2021). Climate Change 2021: The Physical Science Basis. Contribution of Working Group I to the Sixth Assessment Report of the Intergovernmental Panel on Climate Change. Cambridge University Press, Cambridge.

Guillod, B. et al. (2017). A large set of potential past, present and future hydro-meteorological time series for the UK. *Hydrology and Earth Systems Sciences*. doi:10.5194/hess-2017-246.

James, N.D.G. (1990). *The Arboriculturalist's Companion*, 2nd edn. Blackwell Publishers Ltd, Great Britain.

Li, P., Wang, X., Hardy, R.W., Gatlin, D.M. (2004). Nutritional value of fisheries bycatch and by-product meals in the diet of red drum (*Sciaenopsocellatus*). *Aquaculture* 236, 485–496.

Mahmoud, M.K. (2011). *Water Management in Sandy Soil Using scattering Method.* Agricultural Science (Soil Science), Ain Shams University, Egypt.

Mezghani, M.A., MasmoudiCharfia, C., Gouiaa, M., Labidi, F. (2012). Vegetative and reproductive behaviour of some olive tree varieties (Oleaeuropaea L.) under deficit irrigation regimes in semi-arid conditions of Central Tunisia. Scientia Horticulturae 146, 143–152

Pachauri, R.K., Meyer, L.A. *Climate Change 2014: Synthesis Report Contribution of Working Groups I, II and III to the Fifth Assessment Report of the Intergovernmental Panel on Climate Change*. Intergovernmental Panel on Climate Change, Geneva.

Pecl, G. et al. (2017). Biodiversity redistribution under climate change: Impacts on ecosystems and human well-being. *Science* 355/6332, eaai9214. doi:10.1126/science.aai9214.

Rajhi, H., Bardi, A., Dakhli, A., Ardhaoui, S., Sacrafi, N., Bousnina, H., Abichou, M., (2021). Valorization of olive by-products in new biobased toothpaste: Health and sustainability advantages. *Biomass Conversion and Biorefinery*. doi:10.1007/s13399-021-01905-z.

Rajhi H, Bardi A, Ajmi L, Bousnina H, Haq I, Mnif I, Abichou M. (2022). Polygenic olive by products to silage production by anaerobic digestion: A new alternative of sustainable agriculture security and socio-economic advantage impact. *Journal of Oasis Agriculture and Sustainable Development* 4(2), 1–10.

Tibbetts, S.M., Milley, J.E., Lall, S.P. (2006). Apparent protein and energy digestibility of common and alternative feed ingredients by Atlantic cod, Gadusmorhua (Linnaeus, 1758). *Aquaculture* 261, 1314–1327

UNDRR. (2015). *Sendai Framework for Disaster Risk Reduction (2015-2030)*. United Nations Office for Disaster Risk Reduction, Geneva.

URAM. (2010). The Urbanism and International Development Study Office: Master Plan for the Development of the Economic Region of the South-East. URAM

Zhu, C., Liu, Z. (2020). Weakening Atlantic overturning circulation causes South Atlantic salinity pile-up. *Nature Climate Change* 10/11, 998–1003. doi:10.1038/s41558-020-0897-7

# Section B

## Waste Management and Sustainability

# 5 Sustainable Waste Management and Circular Economy Paradigm

*Jean Pierre Doussoulin and Qahraman Kakar*

## 5.1 INTRODUCTION

The rise in world population is negatively affecting our environment. Capitalism has contributed to the rise of the biosphere crisis affecting the functioning of the biosphere and the economy (Southwick, 1996). There is a considerable amount of literature on the negative effect of human consumption practices on long-term biosphere stability (O'Connor, 1991, 1994a). However, the term circular economy has emerged as a controversial issue. Schroeder et al. (2019) argue that if we want to achieve the UN Sustainable Development Goals (SDGs) by 2030, there needs to be enough food, energy and raw materials for the majority of the world's population. Therefore, studies propose a switch from a linear economy toward a circular economy (Esposito et al., 2018; Korhonen et al., 2018a, 2018b; Bonciu, 2014). On the contrary, some authors believe that a variety of business models can extend the life cycle of products on the market. Examples include Walmart, which resells video games after maintenance and thorough inspection, and Phillips, which resells refurbished medical equipment at lower prices (Lacy and Rutqvist, 2015).

A more recent study criticizes the circular economy issue in the biosphere. For example, Millar et al. (2019) claim that the exact relationship between circular economy and environmental deterioration has neither been thoroughly defined nor explored. Therefore, a gap can be observed in the literature that analyzes the contribution of the circular economy to the long-term sustainability of the biosphere. The argument in this context is that the circular economy is not only a concrete political application of industrial ecology but also a useful tool in recycling and repurposing raw materials, promoting energy efficiency and advancing sustainable technology. Integrating the thermodynamics of open systems may counteract resources depletion due to the anthropogenic metabolism. However, these measures may be only a palliative, as they do not eliminate the causes of depletion. Fundamentally, this asymmetrical relationship between the biosphere and the economy is used as a political expression and buzz word in environmental studies (Doussoulin, 2019; Kleinod, 2017).

This chapter attempts to analyze and compare prominent concepts related to biosphere, Leontief theory and Moore Dialectical Method by proposing a new definition of the circular economy. Section 5.2 recalls some controversial issues about the circular economy contributing to biosphere depletion. Section 5.3 briefly discusses Moore's "Four Cheaps" process as a simplified input–output model. Section 5.3 demonstrates how these processes maybe be reinterpreted to express the asymmetrical relationship between the biosphere and the economy.

## 5.2 CIRCULAR ECONOMY ON THE BIOSPHERE

Vernadsky's biosphere is defined from a biogeochemical and a thermodynamic perspective (Vernadsky, 1926). It explains the growth of green plants, autotrophic bacteria and the production of "negative entropy" or neguentropy. Also, Vesnadsky discusses the difference between geochemical

DOI: 10.1201/9781003386902-7

and biogeochemical energies: namely, that geochemical energy measures storage in the geosphere compartments, i.e., atmosphere, lithosphere and hydrosphere but does not account for the energy on the biosphere such as ATP (adenosine triphosphate).

The seminal work of G.E. Hutchinson and Eugène Odum considered the ecosystem a thermodynamic machine capable of maintaining a state of internal order or low entropy (Hutchinson, 1948; Odum, 1971). While our argument is that the literature as mentioned earlier is in line with 'a close the loop' definition of the circular economy. Similarly, Podolinsky states that man profoundly modifies his environment and therein lies the problems of integration of societies in the biosphere. It is on his idea of ecological Marxism that Podolinsky put forward the ideas of a metabolic relationship (Foster and Burkett, 2004). The following three types are widely discussed:

- The eco-socialism theory is supported by many authors including John Bellamy Foster, Richard York, Brett Clark and Paul Burkett, among others. Foster questioned the breakdown of the elemental metabolic or "metabolic rift" between economics and nature (Foster et al., 2010; Moore, 2010a, 2010b, 2017). Burkett (2003, 2004) demonstrated that the formation of the law of value depends not only on the labor force transformation into a commodity but also on other type of labor and the work of nature. However, the work of nature cannot be valued under capitalism; it can only be extracted or mobilized in the service of labor productivity in a strictly capitalist sense (i.e. the production of value).
- The second theory distinguishes the limits of the flow of matter and its metabolism (Erb, 2012; Fischer-Kowalski et al., 2019). This involves quantifying all the material flows involved in the operation of a territory, focusing on direct material consumption (DMC).
- Finally, there is the Common Property Rights Theory (Hardin, 1968). The subject includes contributions from authors such as Frank Knight, Gordon Scott, Anthony Scott, Steven Cheung, Ellinor Ostrom and Gary Libecap. It refers to the condition of environmental degradation, where the majority of the people use a common resource that is characterized by the difficulty of excluding beneficiary and rivalry of use (Ostrom, 1990; Becker and Ostrom, 1995). The tragedy of the commons theory (Hardin, 1968) has been made all too apparent in the last century with the strong growth of the world population and the continuous and growing exploitation of natural resources.

The problem occurs mainly around free resources, defined as those who have access to exhaustible resources. In this scenario, negative product externalities may be *symmetric* or *asymmetric* (Stevenson, 2005). Therefore, the tragedy of the commons is an appropriate starting point for integrating the role of institutions in the good management of natural resources in a neoclassical economic approach.

### 5.2.1 Outlook for a Circular Economy Controversy: Economy Is Not Circular and Cannot Be Circular

Integrating societies in the biosphere needs a holistic approach for studying the cycles of nature, which is as old as the economy. What is at stake in capitalism is its capacity to carry life, referring to the work of Passet, *Economics and the Living* (Passet, 1995). René Passet illustrates the impact of economic activity on the great loops of the biosphere in order to explore this statement. It is therefore necessary to take a holistic approach for ecological economics.

As a general rule, the economy is not circular, but it is entropic (O'Connor, 1991). This has been argued by Haas et al. (2015); their results showed that globally roughly four gigatons per year (Gt/Year) of waste materials are recycled. This flow is of moderate size compared to 62 Gt/Year of processed materials that enter the economy and 41 Gt/yr of outputs (waste). Indeed, economy inputs immediately flows as waste. These wastes may either be landfilled in the biosphere or could be potentially transformed into input using more energy.

### 5.2.2 The Changing Borders between the Biosphere and the Economy: What Are Planetary Boundaries?

What are planetary limits? These are the limits that we should not exceed to avoid a change in the state of the earth system. Also, these limits are the indicators that allow us to understand how and when in the future we will switch toward the Anthropocene. This geological era would have begun with the explosion of our economic activities, the explosion of human demography that has a massive effect on the earth system. This massive effect is precisely measured with these planetary limits (Crutzen, 2006; Steffen et al., 2007).

Furthermore, Rockström et al. (2009), in his famous work published in the journal *Nature*, propose nine planetary boundaries: (i) climate change; (ii) erosion of biodiversity; (iii) consumption of fresh water; (iv) aerosol emissions into the atmosphere; (v) land use and deforestation; (vi) the disruption of biogeochemical cycles of nitrogen and phosphorus; (vii) chemical pollution; (viii) destruction of stratospheric ozone; and (ix) acidification of the oceans. Also, the same article analyzes the disruption of the biogeochemical cycles of nitrogen, and phosphorus, and the effects of the planetary boundaries modification over the climate change and biodiversity erosion. In addition, the literature review extends this planetary boundaries list toward land use and deforestation issues (Steffen et al., 2015). Currently, it has also become more significant to link these limits to COVID-19's emergence and its implications for public health and the regulation of municipal waste produced by urban metabolism (Doussoulin and Colther, 2022; Doussoulin and Mougenot, 2021).

Changes in planetary boundaries can be described through two simultaneous narratives or simultaneous movements. First, the mainstream states that capitalism was able to internalize the living planet and their biogeochemical processes. A second group declares the internalization of capitalism as part of the biosphere (Moore, 2015). In this second group, we can formulate our approach to analyze the relationship between the economy using a complex system composed of biogeochemical elements of natural capital and economic goods. This process must define:

- *Ecological liabilities*: Waste or the ecological footprint of the biosphere cycle elements.
- *Capacity*: The capability of the system to provide resources such as ecosystem services. This feature can be stored or mobilized for the users of the economy. The stored resources remain in the reservoir as the lithosphere or atmosphere Loiret (2013).

Moreover, the differences between use and capacity present a debate between the actors of the economy regarding the storage and the use of natural resources (Hartwick, 1977). This *asset/liability* system (Figure 5.1 left side) can be characterized through four parts: supply, extraction, emissions and fixation. It is important to note that there are more transversal perspectives on the interface between economy and environment, such as bioeconomy, which seeks to provide an alternative to this conjunction through the careful use of natural resources (Mougenot and Doussoulin, 2021).

### 5.2.3 Planetary Boundaries and the World-Systems Analysis

Following the presentation of asset and liability, the system can be adapted to the world-system concept invented by Braudel (1947) and developed by Wallerstein (1987). Wallerstein provided a summary of the path of development of the modern world. It explains the defining characteristics of

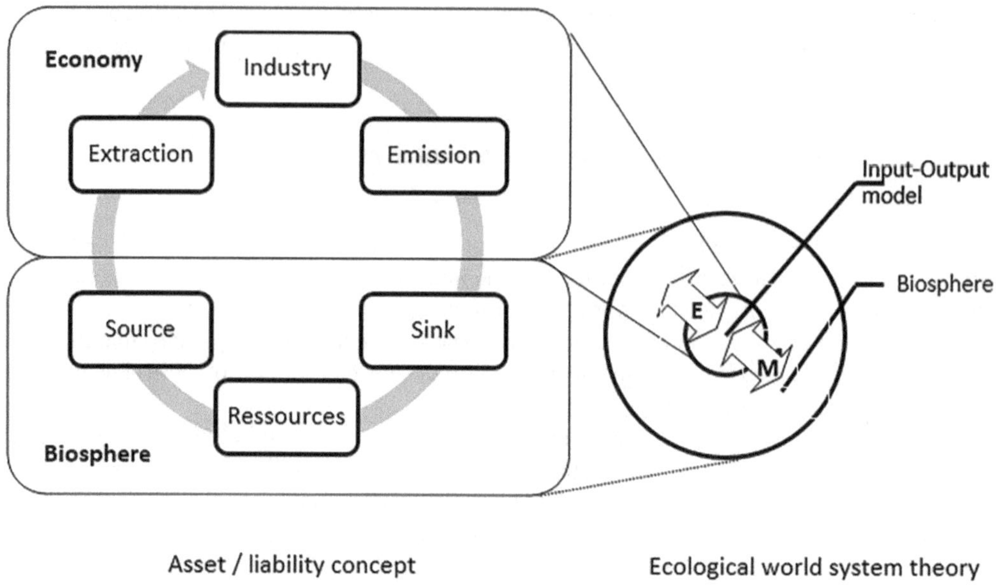


**FIGURE 5.1** Asset/liability concept and world-system theory.

the subject, describing the structures and systems of value exchange on which it is based. In addition, there are three major blocks of intellectual construction in Wallerstein's theory: the school of the Annales, the Marxist Theory and the Dependence Theory.

The Annales school, whose major representative is Braudel (Tanioka, 2001), uses a historical approach to stress long-term social history (Virmani, 2017). Similarly, Wallerstein insists on the long-term approach in line with Braudels. Wallerstein borrows the concept of the centrality of the "accumulation process" and the resulting competitive class struggles from Marx, whereas dependence theory focuses on understanding the periphery by looking at the "center-periphery" relationships. The result of the exchange between the "center" and the "periphery" is a flow of exchange values (Wallerstein, 1987).

The purpose of our model is to present the formalized economic system through *input–output analysis* (Leontief, 1936) or I/O analysis using the Wallerstein structure (Figure 5.1). The periphery is the biosphere, which brings inputs such as biosphere matter and energy to the economy (in terms of natural resource extraction), and leaves the biosphere in terms of pollution or waste. The relationship between conservation of the biosphere and the economy is a complex economic issue. The laws of thermodynamics are relevant because economic activity generates waste. A highly organized system is said to have low entropy (e.g. economic goods); on the contrary, a disorderly system is said to have high entropy (e.g. waste).

There have been several attempts to integrate ecological issues into post-Keynesian models, in order to reconcile these visions. The method implicitly incorporates the laws of thermodynamics to influence the patterns of "flows of funds". In 1971, Georgescu-Roegen made an important distinction between the "flow" resources with regard to energy and matter (e.g. fertilizers and diesel fuel). Thus, resources are not transformed during the agricultural production process (Daly and Farley, 2011; Mayumi et al., 1998).

### 5.2.4 Appropriation and Accumulation Forms

In this section, we will examine the justification of ecological Marxism's discourses. First, we must distinguish two forms of appropriation. The first is the way in which biophysical reproductive processes are used and maintained for resources such as labor, forestry and agriculture. Should we consider humans or the workforce as part of nature or natural biophysical cycles? Moore (2015) makes a series of observations on this subject of the theory of Marx contained in *The Capital* (Marx, 1867). First, Moore argues that humans are part of nature and that their vital activity is a force of nature. Similarly, humans transform the elements contained in the biophysical cycles into the inputs of the economy. This double transformation – of human social relations and of relations within the external nature – implies a third: the continuous transformation of the relationship between humans and the biosphere. Through this transformation, the work produces a variation of the energy that will feed a process of creating economic value. So, there is a work/energy or work/power relationship.

The second corresponds to the geological extraction of matter and energy. In this case, we explore the peak of appropriation of natural resources, which is the reduction of the value of strategic resources for the economic system. Also, this variation reduces the cost of appropriation of natural resource processes by producing the peak in the extraction of petroleum that has been appropriate and accumulated.

Capital works through two strategies of accumulation. One is accumulation by capitalization, centered on the exploitation of the labor force: the production of value within the circuit of capital. The other is accumulation by appropriation, centered on the appropriation of unpaid labor: the unpaid labor of humans and the unpaid labor of nature that reproduces outside of the circuit of capital.

## 5.3 HOLISTIC APPROACH

This regulation system is an expression of the functioning of a subset of human societies and reflects the fundamental "asymmetrical" relations maintained by the capitalist system between the biosphere and the economy. This type of expansion process and biosphere appropriation has been expressed most dramatically in the conquest of the Americas by the European powers. This was achieved through a process of nature "instrumentalization" (Moore, 2010a, 2010b). This mechanism has two stages: the internalization of nature in the value system with waste outsourcing and widening exploitation boundaries.

The current situation requires reconciliation between the social and natural sciences based on a holistic perspective. The solution to the ecosocial controversies that we face requires the use of alternative paradigms based on an integrative vision of the economic system. This approach will clarify the "asymmetry" in the dynamic negotiation of the boundaries between the economy and the environment as described in the famous work, *Is capitalism sustainable?* (O'Connor, 1994b).

The implications of political economics theories and methods can be used to propose new frameworks or analyze long-term changes in natural boundaries. The process of natural capital appropriation uses four basic inputs such as labor, food, energy and raw materials. Moore referred to these four components as "cheap commodities". Synchronized use of the four components manages to generate and sustain a virtuous, productive, dynamic, which generates increased profits. Moreover, the reduction in the synchrony of use of the "cheap commodities" begins to degrade toward an accumulation dynamic. In Moores' words, when "cheap commodities cease to be cheap it begins to be expensive".

### 5.3.1 Previous Input–Output Methods and the Implications for the Measurement of Circular Economy

The input/output or I/O analysis (Leontief, 1936) describes a large set of data that analyzes the interconnectedness of the industries, households and government entities in an area. In addition, it is possible to classify the input–output (I/O) models using different types of monetary or physical units, such as tons of waste. This can be done through a matrix assessing the interdependencies among economic sectors. Usually, the table is presented as a square matrix. The physical flow (energy, matter and waste) are captured via physical input–output matrices.

Indeed, it is possible to distinguish the physical flows according to their origin (source) and destination (sink). A typical presentation of these flows showed in the physical input–output table (PIOT) is a commodity by industry format in physical units, i.e., tons, Mtep. The PIOT is presented using two symmetric tables of inputs and outputs. It can be set up as an inter-industrial material flow (industries by industries flow). Or it can also be expressed by the materials, giving the quantitative relation between the material use (material by material flow) (Pedersen and de Haan, 2006). These flows ensure that the First Law of Thermodynamics is satisfied. According to this law, matter or energy cannot be created or destroyed. These flows also considered the Second Law of Thermodynamics, which implies that the construction process has a tendency to increase the entropy in the system (e.g. geosphere). For example, Daly (1968) proposed four quadrants considering the interaction flows within the economy, flows from the environment to the economy, flows from the economy to the environment and flows in the biochemical cycles.

The PIOT not only shows the physical flow of commodities, but it also allows for waste generation. An example of the applications of PIOT to waste management was proposed by the research of Faye Duchin (1994, 2009). Also, Li (2012) suggested an indirect approach of the classical I/O method proposed by Leontief, which examines the change in material flow through the circular economy issue. Keeping this in view, there is a need to understand the concept of circular economy through the study of their material supply. In this sense, we have an available data source on material consumption, which integrates economic and physical data.

Recently, the model has been expanded. The study of waste flow in a waste management system that applies the principles of the circular economy, using Waste Input-Output Analysis (WIO), was proposed by de Nakamura and Kondo (2002). Other authors, such as Takase, have studied waste management through a life cycle analysis (Takase et al., 2005) Also, Li (2012) and Doussoulin and Bittencourt (2018, 2022) suggested an approach of the "waste input–output" using circular economic indicators. The use of I/O models allows the creation of a formal framework in the analysis and the reflections of four commodities and waste flow into the biosphere.

### 5.3.2 Instrumentalization Mechanism: I/O Background

The use of I/O models provides a formal framework for analyzing and reflecting on the flow of "Four Cheaps" commodities and waste from the biosphere. The general structure of an I/O table shown in Figure 5.1 is divided into: Intermediate demand $x_{ij}$, which represents flows of commodity $i$ with $i = 1\ldots5$; in the processes $j$ with $j=1\ldots5$; these flows are called inter-industry flows or intermediate demand. Final demand $Y_i$ contains data of final demand for the output of each process; final demand is the demand for commodities $i$ with $i=1\ldots5$, which are not used to produce other goods (as opposed to intermediate demand).

Primary inputs $Z_j$ contain inputs used for processes $j$ with $j=1\ldots5$. These are inputs consumed (e.g. labor and foods) and generated (e.g. waste) by the processes of economic systems. The economic system does not directly generate other inputs (e.g. raw materials and energy); these are inputs generated by the biosphere (ecological goods) and consumed by the economic system.

Table 5.1 shows the successive or simultaneous steps to develop three mechanisms using "Four Cheaps" frameworks:

1. The exogenous nature appropriation mechanism considers the internalization of raw material ($x_{11}$) and energy ($x_{21}$) in the economic system using labor power ($x_{31}$) and food ($x_{41}$), the recycling of intermediate waste ($x_{53}$), the outsourcing of ultimate waste ($x_{54}$) into the biosphere and the widening of the economy borders of exploitation of raw material ($x_{15}$) and energy ($x_{25}$) using labor power ($x_{35}$) and food ($x_{45}$).
   - mechanism property states: $x_{11} + x_{21} - x_{53} + x_{54} \leq x_{15}$

**TABLE 5.1**
**I/O General Structure**

| | | | Intermediate Demand | | | | | | |
|---|---|---|---|---|---|---|---|---|---|
| | | | Nature | Nature Property | Recycling | Waste Disposal | Enlarged Bordres | Final Demand | Total Outputs |
| Commodities | Ecological Goods | Raw Raw Material | $X_{11}$ | $X_{12}$ | $X_{13}$ | $X_{14}$ | $X_{15}$ | $Y_1$ | $X_1$ |
| | | Energy | $X_{21}$ | $X_{22}$ | $X_{23}$ | $X_{24}$ | $X_{25}$ | $Y_2$ | $X_2$ |
| | Economical Goods | Labor Power | $X_{31}$ | $X_{32}$ | $X_{33}$ | $X_{34}$ | $X_{35}$ | $Y_3$ | $X_3$ |
| | | Food | $X_{41}$ | $X_{42}$ | $X_{43}$ | $X_{44}$ | $X_{45}$ | $Y_4$ | $X_4$ |
| | | Waste | $X_{51}$ | $X_{52}$ | $X_{53}$ | $X_{54}$ | $X_{55}$ | $Y_5$ | $X_5$ |
| | All Primary Inputs | | $Z_1$ | $Z_2$ | $Z_3$ | $Z_4$ | $Z_5$ | - | - |
| | Total Inputs | | $X_1$ | $X_2$ | $X_3$ | $X_4$ | $X_5$ | | |

2. The endogenous nature management mechanism considers the rights of raw material ($x_{12}$) and energy ($x_{22}$) in the economic system using labor power ($x_{32}$) and food ($x_{42}$), the recycling of intermediate waste ($x_{53}$), the outsourcing of ultimate waste ($x_{54}$) into the biosphere and the widening of the economy borders of exploitation of raw material ($x_{15}$) and energy ($x_{25}$) using labor power ($x_{35}$) and food ($x_{45}$).
   - mechanism property states: $x_{12} + x_{22} - x_{53} + x_{54} \leq x_{15}$
3. The co-option mechanism of the economy considers the rights of raw material ($x_{12}$) and energy ($x_{22}$) into the economic system using labor power ($x_{32}$) and food ($x_{42}$), the recycling of intermediate waste ($x_{53}$) producing an ultimate waste disposal ($x_{54}$) into the economic system.
   - mechanism property states: $x_{12} + x_{22}$- $x_{53} \leq x_{54}$

## 5.4 RELATIONSHIP BETWEEN THE BIOSPHERE AND THE ECONOMY

1. Plunging into the exploratory phase of the nature/society relationship comes with its own difficulties. There is as yet no definite or accepted result for everybody. As we have seen in Jason Moore's historical capitalism, Arrighi (1994) proposes that capitalism has had four long systemic cycles of accumulation since its origins in the 15th century. Each cycle is composed of two phases; in the first phase, a set of private economic actors, in collaboration with state structures, manages to generate and maintain a virtuous productive dynamic, in turn generating increasing profits. Over time, this virtuous dynamic begins to degrade into a financial phase. Thus, the law of diminishing returns begins to apply, and each fraction of invested capital generates a lesser profit. At this point, the systemic cycle of accumulation is entering its financial phase.
2. Each systemic cycle is made up of a center that grows until the moment invested capital generates a lesser profit. Moreover, Arrighi identified "Genova" (which dominated from the 15th to the beginning of the 17th century), the Netherlands (from the end of the 16th to the end of the 18th century), Great Britain (from the mid-18th century to the middle of the 20th century) and the United States (from the end of the 19th to the present) as a

colonial empire. In addition, according to Arrighi, China is called upon to assert itself as the new hegemon in the 21st century (Arrighi, 1994).

### 5.4.1 Capitalist System Configuration

Arrighi's and Moore's vision of capitalism can be framed in this new configuration in the "fabric of life" that can be associated with change in the concept of "oikeios" or set of goods and related men to the same place of residence and production. At the heart of this new definition lives the central idea of a nature that would be external to human relationships, which would be pure, unaltered or healthy and independent of the desires of human beings. This can be represented by the ecological goods of an exogenous nature appropriation mechanism, returning to the vision of Francois Braudel and Immanuel Wallerstein on the debate of the center and the periphery. Speaking of this elongated definition of "oikeios", the system is composed of multiple ecosystems from the center to the periphery. Also, men in the center constantly use their own environment, produce at the same time their own structures of governance power over borders and production (and especially the structures of reproduction of capital).

Inside the center of our system, capitalism considers human labor as a good or resource that is within the trading system. Moreover, in the Marxist conception, the worker works 4 h to receive his own wages as the remuneration of the factors of production and 4 h for the capitalist. In this way, the problem of capitalism today is that the opportunities for the appropriation of goods of production in the zone of exploitation are dramatically reduced, even as the mass of capital traveling the world in search for places to invest is growing. The regulatory system is not able to control the borders of investment and ownership of ecological capital. Such that, capitalism begins to go beyond or broaden its domain to the biosphere that can be represented by "enlarged borders" (Table 5.1).

Foster in their manuscript *Marx's Ecology: Materialism and Nature* proposed two parallel paths in the enlargement of capitalism. On the one hand, the world is at an end: it is the idea of planetary apocalypse or ecological exhaustion. On the other hand, the idea is that capitalism is increasingly confronted with problems of under-consumption of resources in some parts of the world and of social inequality in the distribution of resources and products. To give solutions, we must integrate the two dimensions of the subject, integrating ecology with the analysis of social inequality. However, according to Moore (2015), capitalism was measured in one dimension. That is to say, from the productivity of labor within monetary relations, this phenomenon reduces and simplifies human work and devalues nature's value through the concept of cheap nature. We are not discussing employment or the need to feel useful in nature as a human need in this discussion (Barrera et al., 2022). In this sense, the appropriation of a work and the cheap nature is an integral part of the procedure of capitalism. This procedure can be analyzed further into the endogenous nature management mechanism using Figure 5.1.

### 5.4.2 The Limits of Capitalism in Nature and the Circular Economy Paradigm

As mentioned in the beginning of the chapter, we are witnessing a change in the economic and environmental boundaries. Given that our economic and productive systems have increased their consumption in recent centuries and that we have generated a serious impact on the bio-geophysical system, we are talking about a new geological era called anthropogeny by Crutzen and Steffen (2003) and Steffen et al. (2007). Moreover, we can distinguish two uses of this term. The first corresponds to the way in which the concept of anthropogeny is used in a discussion intended to be cultural. How do humans fit into the web of life? The second important use corresponds to how capitalism has largely increased its capacity to transform the environment in its post-industrial model. In a historical sense, however, these are social and historical dynamics that appear in the 15th, 16th and 17th centuries through the way landscapes and man have transformed themselves into a factor of production.

Serious effects mark the anthropogeny as a new era on earth and the extent of damage has been introduced in the co-option mechanism that considers the appropriation of raw material ($x_{12}$) and energy ($x_{22}$) into the economic system using labor power ($x_{32}$) and food ($x_{42}$), the recycling of intermediate waste ($x_{53}$) producing an ultimate waste disposal ($x_{54}$) into the economic system and into the planet (refer to Table 5.1). The deterioration of the various natural resources and the loss of biodiversity are reflections of this reality. According to the Millennium Ecosystem Assessment (MA) report over the past 50 years, humans have changed ecosystems more rapidly and more profoundly than any comparable period in human history (Millennium Ecosystem Assessment, 2005). These changes in ecosystems have contributed to substantial net gains in terms of welfare and human economic development, but these gains have been made at the increasing costs of the degradation of other ecosystem services and exacerbation of poverty for certain population groups (Büchs and Koch, 2019; Carpenter et al., 2009; Muraca, 2012).

To put it in other way, the process of evolution of cultural systems in recent centuries, especially after the Industrial Revolution, has changed many aspects of the metabolism between society and nature, such that social-cultural evolution has become an evolving macro force (Gowdy, 1997). This type of coevolution process between the ecological system and the social-biophysical, economic and cultural system has been ignored by scientific and economic thought (Gowdy, 1994; Norgaard, 1994; Gual et al., 2010). The challenge of reversing ecosystem degradation while responding to growing demands for their services may be partially met in some cases contemplated by the assessment, but it requires significant changes in policies, institutions and practices not implemented yet (MA, 2005); most of its changes require further studies and analysis of the frameworks, based on an inclusive view of the whole system. Moreover, the rejection of the study of coevolutionary processes and socio-ecological relationships in the consolidated crises perfectly underpinned Cartesian rationalism and the western scientific paradigm.

This paradigm of simplification is based on disjunction, reduction and abstraction (Morin and Weinmann, 1994), with its principles of interdisciplinary and peculiarity, great scientific advances in different fields. But, at the same time, it gave rise to a partial and limited reconciliation of reality; among others, it is divided into three major areas of scientific knowledge: physics, biology and the science of man (Morin, 1977). Thus, all relationships and feedbacks between social and biophysical subsystems require an interdisciplinary approach. However, it has been treated only on mono-disciplinary frameworks of study and analysis. In other words, deliberate progress with the goal of long-term sustainability depends on understanding the dynamics that exist between social and ecological systems (Cumming and Collier, 2005). Therefore, the approach to ecosystem and natural resource management should not only focus on the components of the system but also on their relationships, interactions and feedbacks.

## 5.5. SUMMARY AND CONCLUSIONS

We presented the relations between the biosphere and the economy in a Marxist vision. This system is the expression of the fundamentally asymmetrical relations maintained in the capitalist system between the biosphere and the economy. This paper analyzes not only the process of expansion and appropriation of the biosphere but also includes historical capitalism, focusing on the conquest of the Americas for the European powers. This type of natural capital appropriation process uses four basic inputs or cheap resources such as labor, food, energy and raw materials. Synchronized use of the four components manages to generate and sustain a virtuous productive dynamic, which generates increased profits. Similarly, this chapter analyzes the systemic cycles of accumulation as per Arrighi. This vision made it possible to give a configuration of the capitalist system in the society where web of life may be associated with change in the concept of "oikeios". At the heart of this new definition lies the central idea of a nature that would be external to human relationships, which would be pure, unaltered or healthy and independent of the desires of the human being.

To summarize, the last century's trend has been to dissolve the borders between each cheap commodity to remain in a phase of accumulation, particularly between food, energy and raw materials, which have become increasingly important. This tendency becomes more complicated in the link between food/labor because food and cheap labor are both determined by price and cost of raw materials of agricultural processes that are not highly interchangeable. Another instance of this dissolution of the frontiers between cheap commodities is in the new phase of financialization after the year 2000. In this future, global commodity markets have become digitalized in financial terms, increasing the volatility of their prices.

In other words, this analysis presents the mechanism of instrumentalization, by considering its two stages: the internalization of nature in the value system, and waste outsourcing and expanding the boundaries of exploitation. The mechanism claims the concept of Marx's use value, which integrates nature into the capitalist system of production and "instrumentalization" of the ecosystem processes that have, until now, been autonomous. In this sense, the circular economy's paradigm functions simultaneously as part of the system of "instrumentalization" of the environment and as part of the system by which nature is protected. Creating a waste treatment and recovery process can provide an income for the owners of the process. So, the consumer society is instrumentalizing nature, using natural inputs, throwing and recycling process waste to generate a surplus.

In this regard, the problem with capitalism today is that the appropriation opportunities of the productive assets, which are close to their borders in the exploitation zone, are drastically reduced in the future. As a result, the regulatory system will not be able to control the borders of investment or the appropriation of ecological capital. In this context, the circular economy is, on the one hand, a concrete and political application of industrial ecology. This furthers the need to recycle and re-use material, and improve energy efficiency technologies. On the other hand, it seems appropriate to integrate the thermodynamics vision of open systems, which suggests that there has been some resource depletion due to anthropogenic metabolism. Keeping this in view, it becomes apparent that circular economy action is, in the best case, only a palliative response, which cannot eliminate the cause of depletion.

## ACKNOWLEDGMENTS

We express our gratitude to the National Research and Development Agency of Chile (ANID) for providing the following funds to assist us in conducting this study: Fondecyt- Iniciación Project "Institutional and individual factors to promote responsible behavior in Chile's household waste governance" (projet number 11240954)

## REFERENCES

Arrighi, G., 1994. *The Long Twentieth Century: Money, Power, and the Origins of Our Times*. Verso, London.

Becker, C.D., Ostrom, E., 1995. Human ecology and resource sustainability: The importance of institutional diversity. *Annu. Rev. Ecol. Syst.* 26, 113–133. doi:10.1146/annurev.es.26.110195.000553

Bonciu, F., 2014. The European economy: From a linear to a circular economy. *Romanian J. Eur. Aff.* 14, 78–91.

Braudel, F., 1947. Conflits et refus de civilisation: Espagnols et morisques au xvi e siècle. In: *Annales. Histoire, Sciences Sociales*. Cambridge University Press, Cambridge, pp. 397–410.

Büchs, M., Koch, M., 2019. Challenges for the degrowth transition: The debate about wellbeing. *Futures* 105, 155–165.

Burkett, P., 2003. The value problem in ecological economics: Lessons from the physiocrats and marx. *Organ. Environ.* 16, 137–167.

Burkett, P., 2004. Marx's reproduction schemes and the environment. *Ecol. Econ.* 49, 457–467.

Carpenter, S.R., Mooney, H.A., Agard, J., Capistrano, D., DeFries, R.S., Díaz, S., Dietz, T., Duraiappah, A.K., Oteng-Yeboah, A., Pereira, H.M., 2009. Science for managing ecosystem services: Beyond the millennium ecosystem assessment. *Proc. Natl. Acad. Sci.* 106, 1305–1312.

Crutzen, P.J., 2006. The anthropocene. In: *Earth System Science in the Anthropocene*. Springer, Berlin, pp. 13–18. doi:10.1007/3-540-26590-2_3

Crutzen, P.J., Steffen, W., 2003. How long have we been in the Anthropocene era? *Clim. Change* 61, 251–257.

Cumming, G.S., Collier, J., 2005. Change and identity in complex systems. *Ecol. Soc.* 10, 29.

Daly, H.E., 1968. On economics as a life science. *J. Polit. Econ.* 76, 392–406.

Daly, H.E., Farley, J., 2011. *Ecological Economics: Principles and Applications*. Island Press, Washington, DC.

de Valle Barrera, M., Belloy, P., Mougenot, B., Doussoulin, J.P., 2022. The international impact of Manfred Max-Neef's scholarship: A bibliometric approach. *Int. J. Sustain. Dev.* 25, 1–29. doi:10.1504/IJSD.2022.126458

Doussoulin, J.P., 2019. A paradigm of the circular economy: The end of cheap nature? *Energy Ecol. Environ.* 1–10. doi:10.1007/s40974-019-00145-2

Doussoulin, J.P., Bittencourt, M., 2018. Analysing the circular economy opportunities in the French construction sector related to the sustainable supply chain: A waste input-output analysis. *Int. J. Supply Chain Oper. Resil.* 3, 143–162. doi:10.1504/IJSCOR.2018.090769

Doussoulin, J.P., Bittencourt, M., 2022. How effective is the construction sector in promoting the circular economy in Brazil and France? A waste input-output analysis. *Struct. Change Econ. Dyn.* 60, 47–58. doi:10.1016/j.strueco.2021.10.009

Doussoulin, J.P., Colther, C., 2022. Evaluating the efficiency of municipal solid waste collection services in developing countries: The case of Chile. *Sustainability* 14. doi:10.3390/su142315887

Doussoulin, J.P., Mougenot, B., 2021. The economic, climate change and public health edges of the geopolitics of COVID-19: An exploratory bibliometric analysis. In: *Biotechnology to Combat COVID-19*. IntechOpen. doi:10.5772/intechopen.96797

Duchin, F., 1994. Input-output analysis and industrial ecology. Green. Ind. Ecosyst. 23, 61–68.

Duchin, F., 2009. Input-output economics and material flows. In: *Handbook of Input-Output Economics in Industrial Ecology*. Springer, New York, pp. 23–41.

Erb, K.-H., 2012. How a socio-ecological metabolism approach can help to advance our understanding of changes in land-use intensity. *Ecol. Econ.* 76, 8–14. doi:10.1016/j.ecolecon.2012.02.005

Esposito, M., Tse, T., Soufani, K., 2018. Introducing a circular economy: New thinking with new managerial and policy implications. *Calif. Manage. Rev.* 60, 5–19. doi:10.1177/0008125618764691

Fischer-Kowalski, M., Rovenskaya, E., Krausmann, F., Pallua, I., Neill, J.R.M., 2019. Energy transitions and social revolutions. *Technol. Forecast. Soc. Change* 138, 69–77. doi:10.1016/j.techfore.2018.08.010

Foster, J.B., Burkett, P., 2004. Ecological economics and classical marxism: The "Podolinsky Business" reconsidered. *Organ. Environ.* 17, 32–60. doi:10.1177/1086026603262091

Foster, J.B., Clark, B., York, R., 2010. *The Ecological Rift*. NYU Press, New York.

Gowdy, J.M., 1997. The value of biodiversity: Markets, society, and ecosystems. Land Econ. 73, 25–41.

Haas, W., Krausmann, F., Wiedenhofer, D., Heinz, M., 2015. How circular is the global economy?: An assessment of material flows, waste production, and recycling in the european union and the world in 2005. *J. Ind. Ecol.* 19, 765–777. doi:10.1111/jiec.12244

Hardin, G., 1968. Tragedy of the commons. *Nature* 162, 1243–1258.

Hartwick, J.M., 1977. Intergenerational equity and the investing of rents from exhaustible resources. *Am. Econ. Rev.* 67, 972–974.

Hutchinson, G.E., 1948. Circular causal systems in ecology. *Ann. N. Y. Acad. Sci.* 50, 221–246.

Kleinod, M., 2017. *The Recreational Frontier: Ecotourism in Laos as Ecorational Instrumentality*. Universitätsverlag Göttingen, Göttingen.

Korhonen, J., Honkasalo, A., Seppälä, J., 2018a. Circular economy: The concept and its limitations. *Ecol. Econ.* 143, 37–46.

Korhonen, J., Nuur, C., Feldmann, A., Birkie, S.E., 2018b. Circular economy as an essentially contested concept. *J. Clean. Prod.* 175, 544–552.

Lacy, P., Rutqvist, J., 2015. *Waste to Wealth: The Circular Economy Advantage*. Springer, New York.

Leontief, W.W., 1936. Quantitative input and output relations in the economic systems of the United States. *Rev. Econ. Stat.* 18, 105–125.

Li, S., 2012. The research on quantitative evaluation of circular economy based on waste input-output analysis. *Procedia Environ. Sci.* 12, 65–71.

Marx, K., 1867. *Capital, Vol. 1: A Critique of Political Economy*. N. Y. Vintage, New York.

Mayumi, K., Giampietro, M., Gowdy, J.M., 1998. Georgescu-Roegen/daly versus solow/stiglitz revisited. *Ecol. Econ.* 27, 115–117.

Millar, N., Mclaughlin, E., Börger, T., 2019. The circular economy: Swings and roundabouts? *Ecol. Econ.* 158, 11–19.

Millennium Ecosystem Assessment, 2005. *Ecosystems and Human Well-Being*. Island Press Washington, DC:

Moore, J., 2015. *Capitalism in the Web of Life : Ecology and the Accumulation of Capital*. Verso, New York.

Moore, J., 2017. Metabolic rift or metabolic shift? dialectics, nature, and the world-historical method. *Theory Soc.* 46, 285–318.

Moore, J.W., 2010a. Amsterdam is standing on Norway' part I: The Alchemy of capital, empire and nature in the diaspora of silver, 1545–1648. *J. Agrar. Change* 10, 33–68.

Moore, J.W., 2010b. Amsterdam is standing on Norway' part II: The global north atlantic in the ecological revolution of the long seventeenth century. *J. Agrar. Change* 10, 188–227.

Morin, E., 1977. La Nature de la Nature. Seuil, Paris.

Morin, E., Weinmann, H., 1994. La Complexité Humaine. Flammarion, Paris.

Mougenot, B., Doussoulin, J.-P., 2021. Conceptual evolution of the bioeconomy: A bibliometric analysis. *Environ. Dev. Sustain.* 1–17. doi:10.1007/s10668-021-01481-2

Muraca, B., 2012. Towards a fair degrowth-society: Justice and the right to a 'good life' beyond growth. *Futures* 44, 535–545.

Nakamura, S., Kondo, Y., 2002. Input-output analysis of waste management. *J. Ind. Ecol.* 6, 39–63.

O'Connor, M., 1991. Entropy, structure, and organisational change. *Ecol. Econ.* 3, 95–122. doi:10.1016/0921-8009(91)90012-4

O'Connor, M., 1994a. The second contradiction of capitalism. The material/communal conditions of life. *Capital. Nat. Soc.* doi:10.1080/10455759409358613

O'Connor, M., 1994b. *Is Capitalism Sustainable ? : Political Economy and the Politics of Ecology, Democracy and Ecology*. Guilford Press, New York. doi:10.1177/108602669600900420

Odum, E., 1971. Fundamentals of ecology, 3ème éd. Saunders, PA.

Ostrom, E., 1990. *Governing the Commons: The Evolution of Institutions for Collective Action*. Cambridge University Press, Cambridge.

Passet, R., 1995. L'économique et le vivant. doi:10.3917/econo.passe.1996.01

Pedersen, O.G., de Haan, M., 2006. The system of environmental and economic accounts-2003 and the economic relevance of physical flow accounting. *J. Ind. Ecol.* 10, 19–42.

Rockström, J., Steffen, W., Noone, K., Persson, Å., Chapin III, F.S., Lambin, E.F., Lenton, T.M., Scheffer, M., Folke, C., Schellnhuber, H.J., 2009. A safe operating space for humanity. *Nature* 461, 472.

Schroeder, P., Anggraeni, K., Weber, U., 2019. The relevance of circular economy practices to the sustainable development goals. *J. Ind. Ecol.* 23, 77–95.

Southwick, C.H., 1996. Global Ecology in Human Perspective. Oxford University Press, Oxford.

Steffen, W., Crutzen, P.J., McNeill, J.R., 2007. The Anthropocene: Are humans now overwhelming the great forces of nature. *AMBIO J. Hum. Environ.* 36, 614–622.

Steffen, W., Richardson, K., Rockström, J., Cornell, S.E., Fetzer, I., Bennett, E.M., Biggs, R., Carpenter, S.R., De Vries, W., De Wit, C.A., 2015. Planetary boundaries: Guiding human development on a changing planet. *Science* 347, 1259855.

Stevenson, G.G., 2005. *Common Property Economics: A General Theory and Land Use Applications*. Cambridge University Press, Cambridge.

Takase, K., Kondo, Y., Washizu, A., 2005. An analysis of sustainable consumption by the waste input-output model. *J. Ind. Ecol.* 9, 201–219.

Tanioka, T., 2001. Fernand Braudel and the "Annales" *School. Hum. Geogr.* 53, 327–344. doi:10.4200/jjhg1948.53.327

Vernadsky, V.I., 1926. The Biosphere. Leningrad, Nauchtekhizdat (in Russian).

Virmani, A., 2017. Social responsibility of the historian: The Annales' agenda in perspective. *Econ. Polit. Wkly.* 52, 46–53.

Wallerstein, I., 1987. *World-Systems Analysis*. Stanford, CA.

# 6 Waste as a Source of Raw Material in Circular Economy for Value-Added Products

*Simone Kubeneck, Caroline Dalastra, Suzana Fátima Bazoti, Júlia Nerling, Gabriel Henrique Klein, and Helen Treichel*

## 6.1 INTRODUCTION

Due to population growth, the increase in demands from various economic and industrial sectors resulted – and still results – in the high generation of waste. The agro-industrial and food sectors are among the most prominent waste generators. In Brazil alone, the production of food such as fruits, animal protein, and vegetables, among others, reaches approximately 11,850 tons (FAO 2022). Therefore, ways to reuse waste from these sectors is a viable alternative for sustainable development and application of the circular economy concept, making these wastes a raw material and creating new economic opportunities by reusing them properly to obtain added value bioproducts (Teigiserova et al. 2020; Chavan et al. 2022).

The meatpacking industries, belonging to the agro-industrial sector and responsible for animal protein processing, have great waste generation, mainly in the slaughtering sector, where there is the generation of waste such as swine hair and chicken feathers, hooves, beaks, and blood, among others. These mostly have landfills and incineration as their final destination, which can lead to serious environmental problems such as reduced oxygen and lack of soil eco-functionality, as well as the appearance of microorganisms and disease-causing vectors (Dedeke et al. 2016; Ramires et al. 2020). In addition, chicken feathers and swine hair are residues that have keratin in high amounts in their composition. Due to this, they are considered natural polymers and, consequently, are difficult to degrade by the action of microorganisms and proteases such as pepsin, trypsin, and papain (Papadopoulos 1986; Gupta & Ramnani 2006). However, due to the composition rich in proteins such as keratin and components such as carbon and nitrogen, these residues become raw materials for producing value-added products.

As far as the agro-industrial sector, the food sector also has a large generation of waste in its production process. For example, in industries responsible for fruit processing, residues such as bagasse and peels are generated and fruits that are unsuitable for human consumption. In addition to food residues, cereals, and sugarcane bagasse, among other residues from agriculture, are considered viable biomass for the production of bioenergy due to the availability of fermentable sugars such as glucose, fructose, and sucrose, and polymers such as cellulose, hemicellulose, and lignin that can be converted into bioproducts more economically and effectively (Srivastava et al. 2021; Torre et al. 2017).

In the current scenario, efforts are being made to apply the concept of a circular economy that involves recovering, recycling, and reconditioning of waste such as those from agro-industrial and food processes, converting them into value-added products such as biofuels, biofertilizers, enzymes, chemicals, and biopolymers. Numerous technological methods involve the steps that precede the

DOI: 10.1201/9781003386902-8

conception of these bioproducts, from pretreatment to processes involving microbial fermentation and enzymatic hydrolysis. Therefore, microorganisms play a fundamental role since they produce industrially essential enzymes (such as cellulase, protease, and amylase) that can increase the biodegradation of waste to modify its structure and consequently, carry out the production of bioproducts in an economically and environmentally sustainable way as these are low-cost processes with minimal generation of polluting waste.

In this sense, this chapter addresses the most recent studies on the potential of agro-industrial and food residues to obtain value-added products and the different technologies involved in the process, discussing future perspectives and the main challenges applied.

## 6.2 IMPORTANCE OF PRETREATMENT OF WASTE

Agro-industrial waste is generated abundantly and comes from various sources, such as agriculture, food processing, domestic waste, industry, and commerce. The leading destination is the sanitary landfill; however, the inadequate disposal of these residues can compromise the environment through the release of methane, one of the greenhouse gases that contributes to the climate changes that are being faced (Nair et al. 2022). These residues, mostly of biological basis, are formed by compounds that can be precursors of new value-added products. Through the circular economy, it is possible to transform these raw materials, also called biomass, into biofuels, bioenergy, and differentiated bioactive compounds. A promising strategy is the pretreatment of these materials, which makes the components present in their structure available to be used by microorganisms in biochemical reactions forming new compounds, such as enzymes, antibiotics, phytochemicals, biofertilizers, and biofuels, among others (Yaashikaa et al. 2022).

Effective pretreatment of biomass is an essential step in biorefinery, as it can guarantee the commercial viability of the bioprocess. Among its primary functions, we highlight the greater accessibility to the compounds in the lignocellulosic matrix, improvement in biocompatibility, decrease in cellulose crystallinity, and increase in product yield, aiming to improve the biomass conversion capacity. Pretreatments can be physical, chemical, or biological. Its main objective is to modify the physical structure of the biomass by increasing the surface area with the decrease of the particles and improving the mass transfer in the subsequent processes (Zhou & Tian 202; Treichel et al. 2020). Among the physicists, milling, grinding, ultrasonic vibration, microwaves, and heat stand out.

Grinding and extrusion are attractive processes widely used due to their simplicity and can be used previously or in combination with other pretreatments, such as biological ones. Its most significant benefits are the absence of chemical reagents and, consequently, the decrease in the formation of inhibitor compounds for the later stages (Treichel et al. 2020).

Steam explosion is also an economical technique widely used to treat biomasses with higher recalcitrance, such as hardwood. It has the potential for numerous products, low monosaccharide degradation, and lower energy consumption. To this process, acids or bases can also be added, characterizing the acid or alkaline steam explosion (Zhou & Tian 2022; Treichel et al. 2020).

Ultrasound performs the process through cavitation and is also low-cost, relatively easy to operate, and can treat several samples simultaneously. Furthermore, compared with traditional heating, this technique has low degradation of sugars and formation of byproducts (Rahmani et al. 2022; Treichel et al. 2020). Likewise, microwave irradiation, which consists of a dielectric heating mode that can biodegrade biomass by altering its internal structure, has been extensively studied in the last three decades, presents promising results and there is already research on a pilot scale.

Chemical pretreatments physically and chemically modify the biomass structure through reagents, which promote the cleavage of covalent cross-links in the lignocellulosic matrix. They reduce crystallinity; increase the degradation of cellulose, hemicellulose, and lignin; and have high reaction rates and greater energy efficiency. However, they can form undesirable intermediate products, mainly for subsequent biological processes and loss of fermentable sugars (Rahmani et al. 2022). Among the chemical methods, the most used are acid or alkaline pretreatments. However, in addition to these, several studies report using organic solvents such as maleic, succinic, levulinic, acetic, and oxalic acids and using low-boiling alcohols such as methanol, ethanol, and glycerol. Ionic liquids, deep eutectic solvents, and ozonation have also shown promising results (Zhou & Tian 2022; Rahmani et al. 2022).

The most frequently used acids are inorganic acids, such as sulfuric acid, hydrochloric acid, and nitric acid, due to their low cost, high ability to change the crystallinity of cellulose and weaken the hydrogen bonds between cellulose and glucose, reducing the recalcitrance of biomass (Jamaldheen et al. 2022). The acid process is easy to operate and presents a higher degree of hemicellulose solubilization, benefiting subsequent hydrolysis. The use of dilute sulfuric acid, for example, facilitated the diffusion of enzymes to degrade the carbohydrates present in wheat straw (Rahmani et al. 2022). The choice of the acid medium also provides good yields of sugars in enzymatic saccharification.

The disadvantages of using the acid route are forming inhibitors, such as furfural and hydroxymethylfurfural, difficulty removing acids after hydrolysis, relatively high waste generation, high cost, and equipment corrosion. Furthermore, if the conditions used are severe, they may interfere with the bioconversion of lignocellulosic biomass due to the formation of pseudo-lignin derived from furans (Jamaldheen et al. 2022). To reduce the generation of inhibitory compounds, there are studies combining acid pretreatment with biologicals resulting in higher glucose and ethanol yields (Meenakshisundaram et al. 2021).

Alkaline pretreatment makes the use of bases to treat the biomass. The most commonly used are sodium hydroxide, potassium hydroxide, ammonium hydroxide, and calcium hydroxide. Through hydroxyl radicals, they can separate hemicellulose and lignin and slightly dissolve cellulose and hemicellulose, breaking the hydrogen bond between these two compounds and promoting the solubilization of acetyl groups (Rahmani et al. 2022; Treichel et al. 2020). In this way, the alkaline medium can change the lignocellulosic structure, first breaking the lignin, then breaking the hemicellulose, and, finally, breaking the cellulose polymorphic crystal lattice. Thus, it solubilizes lignin, which contributes to faster hydrolysis and saccharification. Changing the process parameters, such as alkali concentration and temperature, can achieve different degrees of delignification (Jamaldheen et al. 2022). Among the advantages of the alkaline route, it is possible to mention the lower formation of inhibitors. In addition, it is more economical due to the project's simplicity; the operation takes place at milder temperatures, preserves the structure of the carbohydrates, has good synergy, and can be associated with other processes, mainly biotechnological ones (Rahmani et al. 2022). However, the alkaline treatment needs a longer retention time than the other reactors with greater capacity, which implies an increase in costs. In addition, it can form byproducts from lignin, such as vanillin and syringaldehyde polymers, which are harmful in biological processes and make it difficult to recycle or reuse the final digested substrate (Rahmani et al. 2022).

The technique of integrating pretreatments consists of the sum of two or more methods that, when together, can result in the best use of biomass. In general, this integration aims to overcome the compromised efficiency of unmatched pretreatments (Treichel et al. 2020). It is worth mentioning that the method chosen to pre-treat needs to be selected to guarantee the technical–economic feasibility and, mainly, the efficiency in the downstream operations, that is, the excellent choice of the sequence that will be used in the treatment of the biomass is fundamental, as it will significantly influence the outcome of the treatment (Jamaldheen et al. 2022).

The most used alkaline pretreatment process is with sodium hydroxide. Its use results in increased surface area and more excellent enzymatic hydrolysis, caused by the delignification reaction and decrease in cellulose crystallinity. This pretreatment is carried out at average temperatures of 100°C to ensure satisfactory results. However, as already mentioned, there is a possibility of facilitating and/or improving the outcomes by combining more than one pretreatment. In this case, it is appropriate to integrate biological pretreatment since it can enhance delignification and lower the temperature of the alkaline treatment, thus reducing heating costs in the process (Meenakshisundaram et al. 2021).

For this integration, in most cases, biological pretreatment is used before alkaline pretreatment. This was made possible by the combination resulting in better cellulose digestibility and a reduction in the production of toxic inhibitors to subsequent processes (Meenakshisundaram et al. 2021). Therefore, this process proves to be both more economical and more efficient. Biological pretreatment occurs with the inclusion of microorganisms and/or bio-based products to decompose the biomass. In mild conditions, they have considered ecologically correct processes since there is no need to insert chemical material in the development of pretreatment (Nair et al. 2022).

The methods to biologically pre-treat can include pretreatment by anaerobic digestion, microbial pretreatment, and enzymatic pretreatment, among others (Zhou & Tian 2022). In the anaerobic technique, digestion occurs without oxygen to produce products from organic matter such as those from agricultural practices (Yaashikaa et al. 2022). It is worth mentioning the great importance of this process due to the valorization of farming residues because it is economical and simple and can generate biogas that can be widely used at the end of the process.

Microbial pretreatment is an economical method as it consumes little energy, in addition to being able to degrade from 12% to 20% of the lignin and hemicellulose present in the substrate (Rahmani et al. 2022). However, this process causes the accumulation of mycelial mats in the pre-treated biomass, hindering the accessibility of hydrolytic enzymes to cellulose and the production of biogas (Zhou & Tian 2022). In most fungal pretreatments, white rot fungus, brown rot fungus, and soft fungus are used; however, nothing prevents the use of other microorganisms, as well as the combined use of more than one type of fungus, if the results prove to be efficient.

Considering that the use of pretreatments aims at greater accessibility to the compounds present in residues that have a lignocellulosic matrix without the generation of inhibitors for the fermentation step, enzymatic pretreatments are an alternative since, through the use of enzymes, there is no generation of residues, besides not having the age of inhibitors. Furthermore, it is also worth mentioning that accessory enzymes are often used to provide synergistic effects and greater availability of the components of the biomass structure (Zhou & Tian 2022), which may come from the metabolic action of microorganisms in contact with waste.

## 6.3 ROLE OF MICROORGANISMS AND ENZYMES IN THE CONVERSION OF WASTE INTO VALUE-ADDED PRODUCTS

The use of agro-industrial and food residues aiming at the production of bioproducts (e.g., enzymes, chemicals, and bioenergy) in microbial fermentation processes is a viable alternative due to two main reasons: (i) the wide and easy availability and (ii) the high amounts of nutrients such as carbon, nitrogen, proteins, and minerals, which make fermentation media composed of these residues a suitable environment for microbial growth (Raina et al. 2022; Sadh et al. 2018).

In addition to their use as a fermentation substrate, these residues, as well as the environment where they are discarded, are a promising source for the prospection of microbial strains, which can be used with superior performance to other strains in the bioconversion processes. This is because the fermentation medium that has the same residue as its raw material resembles the microorganism's natural habitat and, therefore, the inhibitory effects from the composition of the residue do not drastically affect the microorganism in the conversion into value-added products (Kumar & Dwivedi 2019; Chilakamarry et al. 2022; Šelo et al. 2021).

Recent studies report that strains of *Aspergillus* sp., *Fusarium oxysporum*, *Bacillus* sp., *Proteus* sp., and *Providencia* sp., among other microorganisms, isolated from agro-industrial residues such as chicken feathers and residues from the production process of the food sector can be used to obtain enzymes and produce products aimed at plant development (Preczeski et al. 2020; Kim et al. 2021). In addition, yeasts capable of assimilating glucose, fructose, and sucrose – essential characteristics in processes aimed at producing biofuels – can also be obtained from agro-industrial residues such as poultry litter and residues such as pieces of fruit (Camargo et al. 2018). About 45 strains could be isolated from poultry litter and fruits, such as jatobá do cerrado (*Hymenaea courbaril* var. *stigonocarpa*), uvaia (*Eugenia pyriformis*), umbu (*Spondias tuberosa Arruda*), acerola (*Malpighia glabra* L.), ubajay (*Hexachlamys edulis*), and pequi (*Caryocar brasiliense*) in the study carried out by Camargo et al. (2018), which showed the ability to assimilate sugars and ferment them, thus showing potential in the bioconversion of sugar-rich residues into biofuels. Residues of apple, pineapple, and onion are also considered promising sources for the prospection of microbial strains, with the potential for the production of cellulose, fructooligosaccharides, and animal feed, using food residues as raw material (Kaur & Shah 2022; Tan et al. 2020; Ng et al. 2020).

The use of agro-industrial and food residues as substrates in microbial fermentation processes, in addition to being sources of nutrients, due to their easy availability, allows their bioconversion into bioproducts economically since it is about the reuse of the residue in bioconversion low processes cost. Furthermore, microbial strains appear to be an essential part of the process of bioconversion of waste since the production of value-added products by fermentation occurs through the action of microorganisms (mainly fungi and bacteria) on these substrates (Bharathiraja et al. 2017). In general, the bioconversion process of waste into value-added products consists of microbial absorption and fermentation mechanisms, followed by the extraction of the desired product (Chavan et al. 2022).

Filamentous fungi are examples of microorganisms used for over a century as versatile cell factories, high producers of enzymes, antibiotics, and organic acids, among other active compounds that are still investigated today (Wösten 2019). Filamentous fungi and yeasts, when in contact with the residue used as substrate in fermentation processes, in the solid state, for example, develop in the form of mycelium and biofilm, respectively (Šelo et al. 2021). Filamentous fungi are formed by hyphae, which, if thick, allow the water in the fermentation medium to move through capillary action, resulting in a biofilm (Šelo et al. 2021). On the other hand, bacteria form microcolonies that attach, producing extracellular cells that bind to other cells through the outer membrane and, together with the water present in the medium, create a fatty layer for biofilm formation (Sonawane et al. 2022)

Biocatalytic processes can be carried out from the formation of biofilms by microorganisms. The biofilm considered mature is developed concomitantly with the dispersion of cells from other biofilms present in the medium. Adherence to the surface of residues in fermentation media makes this biofilm a stable biocatalyst for an extended period (Leonov et al. 2021; Muffler et al. 2014). When applied in fermentations whose raw material is lignocellulosic residues, such as rice bran, biofilms can tolerate high concentrations of inhibitory components from the residues themselves. Due to this fact, they are considered good biocatalysts, resulting in increased productivity and yield in fermentation processes aimed at obtaining bioenergy, such as bioethanol (Leonov et al. 2021; Todhanakasem et al. 2014).

In the agroindustry, residues such as chicken feathers and swine hair are rich in keratin. In its fermentation process, bioconversion begins with fungal degradation, where filamentous fungi produce hyphae that develop around the main constituent of the waste, in this case, the keratin ribbon, later going through processes of surface erosion, lifting of cuticular scales. Finally, the microorganism radial penetrates into the waste (Preczeski et al. 2020; Călin et al. 2017).

After the microbial processes, enzymes and microorganisms act to change the structure of substrates in the bioconversion process. In keratin-rich residues, enzymes such as disulfide reductases work by reducing the disulfide bonds present in the keratin structure, performing enzymatic hydrolysis by endo- and exoproteolytic enzymes, obtaining final product enzymes such as keratinases in addition to amino acids and peptides (Yamamura et al. 2002; Preczeski et al. 2020).

Lignocellulosic residues are mainly composed of cellulose, hemicellulose, and lignin. The main sources of these residues are the primary and secondary products of forest extraction, agriculture, urban residues, and food residues such as fruits, sugarcane bagasse, cereals, and wood processing residues. A set of enzymes act to modify these residues in the bioconversion process. Enzymes such as peroxidase and laccase act by altering the structure of lignocellulosic residues through oxidation reactions capable of breaking the lignin polymer to bioconvert into other value-added products (Cajnko et al. 2021; Tan et al. 2022).

Peroxidase enzymes such as lignin peroxidase and manganese peroxidase act in different ways. Lignin peroxidase oxidizes phenolic and non-phenolic compounds and amines and aromatic ethers. On the other hand, manganese peroxidase oxidizes Mn (II) present in the residue to Mn (III), and this oxidizes phenolic compounds producing phenoxy radicals that act in the depolymerization process (Pérez et al. 2002; Tan et al. 2022). Laccases are another group of enzymes active in the structural modification of lignocellulosic residues. These oxidoreductase enzymes, through mediators, act by degrading polyphenolic and non-phenolic compounds present in the lignin structure (Gonzalo et al. 2016; Tan et al. 2022; Thurston 1994; Cajnko et al. 2021). Cellulases act to hydrolyze β-1,4-glycosidic bonds present in the cellulosic layer. Endoglucanases hydrolyze the internal bonds in amorphous regions of cellulose, causing the release of new terminal ends that are hydrolyzed by cellobiohydrolases, the products obtained through these reactions being the sources of carbon and energy used by microorganisms during fermentation processes (Leschine 1995; Pérez et al. 2002). Furthermore, lignocellulosic residues are composed of hemicellulose and cellulose that also use enzymes to be bioconverted into value-added products.

Finally, the hemicellulose layer of these residues is mainly composed of xylan, and due to this, a set of enzymes act to degrade these compounds. One of the most common forms of hemicellulose, O-acetyl-(4-O methylglucurono)xylane, has its structure modified through four different enzymes such as endoxylanases, acetyl esterase, glucuronidase, and b-xylosidase (Pérez et al. 2002). For the degradation process of O-acetyl-galactoglucomannan, first enzymes such as endomannases break the polymer of the structure. From then on, acetyl glucomannan esterases and α-galactosidases act by removing acetyl groups and galactose residues, and, finally, there is the breaking of β bonds -1,4 generated by endomanases, resulting in sugars and chemical products such as acetic acid at the end of the process (Pérez et al. 2002)

Microbial fermentation processes, as well as enzymatic hydrolysis that occurs in sequence or in conjunction with microbial fermentation, are considered the essential steps in the bioconversion of waste, especially when sustainable production is required in an environmentally friendly way. Through processes such as these, waste is used to obtain bioproducts in a circular economy context, without generating other waste considered pollutants and making waste from the agroindustry and food sector a set of raw materials to get bioproducts.

## 6.4 VALUE-ADDED PRODUCTS OBTAINED FROM WASTE

With the development of biotechnology, waste is no longer an environmental problem as it is adding value to biotechnological processes. Research carried out in recent decades has been directing its efforts toward the use of organic waste in the development of bioproducts and bioenergy, as they are renewable and low-cost resources, thus promoting the minimization of waste disposal. Using them as a source of biological raw material in integrated processes to obtain different bio-based products is an essential and sustainable strategy with several socioeconomic and ecosystem benefits.

These residues come from the food and agro-industrial sectors, which have different chemical compositions and generate other products (see Figure 6.1). Rice processing residues (husks, bran, and grains) comprise about 32% cellulose and 21% lignin. Soybean harvest residues have about 50% cellulose, 25% hemicellulose, and 25% lignin. Waste from the food industry is rich in carbohydrates, lipids, proteins, vitamins, and minerals. Cereal and wheat bran residues have high cellulose, hemicellulose, and starch levels. Horns, hooves, feathers, and hair from slaughtered animals are rich in biologically active substances (found in keratin fat). Their appreciation is also paramount (Astolfi et al. 2019; Xu et al. 2020; Sivaramakrishnan et al. 2021; Preczeski et al. 2020).

The basis of the circular economy (reuse, repair, and renew) aims to improve efficiency both in the production and in the use of products, valuing the life cycles and the strategy of using goods and raw materials. Biorefineries, for example, play an essential role in biomass conversion, as they use waste entirely to generate energy and biologically-based chemicals. In addition, solid waste processed in biorefineries is rich in proteins, carbohydrates, and minerals, having a high potential for developing high-value products, such as enzymes, chemicals, biopolymers, biofertilizers, and biofuels (Jain et al. 2022).

The concept of bioenergy is intrinsic to the idea of sustainability through renewable energy generation. Bioenergy accounts for approximately 14% of global energy demand (Singh et al. 2022a). Although most renewable energy comes from wind, solar, tidal, hydroelectric, and geothermal sources, liquid fuels still predominate in this area. Biofuels such as bioethanol, biodiesel, butanol, and biohydrogen are sustainable alternatives for replacing fossils and have been gaining prominence in scientific research (Boro et al. 2022).

First-generation biofuels are produced from sugarcane, beetroot, rice, corn, wheat, and soybean oil. The second generation is already made from renewable resources, such as crop residues used for the first generation, sludge of effluent, and residual cooking oil, among others (Nair et al. 2022). In addition to fuels, agro-industrial waste can be used for energy generation, biocompounds, and biofertilizers. The use of waste is an excellent resource to increase energy production in countries and reduce environmental pollution within a circular economy.

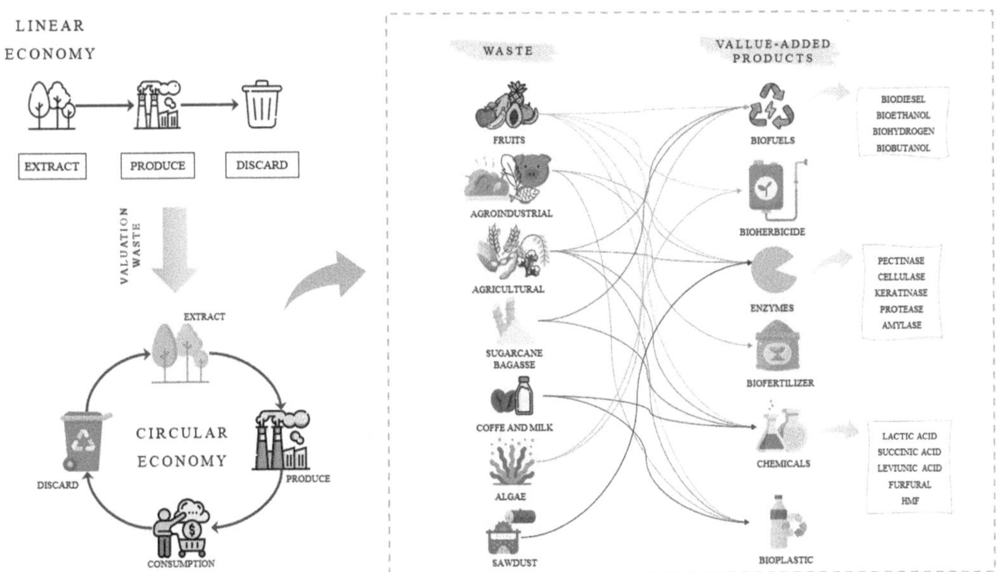


**FIGURE 6.1** Possibilities of obtaining value-added products from waste in the context of the transition from the linear economy to the circular economy. (*Source*: The authors 2022).

### 6.4.1 Biofuels

The use of liquid biofuels has increased considerably in recent decades. At the same time, the production also expanded, reaching a growth of 65% for bioethanol, 25% for biodiesel, and 10% for the others, between 2000 and 2016. However, there is still a significant dependence on oil, and it becomes increasingly essential to find carbon-based alternatives with the power to provide energy constantly in a shorter time. In this sense, biomass used to produce biofuels meets these demands with a wide range of products (Boro et al. 2022). In addition, the production and use of biofuels from agro-industrial waste meet the Sustainable Development Goals, which aim to slow climate change and reduce global warming.

#### 6.4.1.1 Bioethanol

Currently, bioethanol is the primary renewable fuel. Its main route of obtaining is fermenting raw materials based on sugar, such as cornstarch and sugarcane. However, this path competes with the supply of food. With that, the transition from the first-generation to second-generation fuels took place, which has advantages over the first, mainly due to the abundant availability of raw material at low cost, nature not being competitive with food-based crops, being an excellent renewable resource (Singh et al. 2022b). In 2020 alone, the world ethanol market was valued at US$93.7 billion and is estimated to exceed US$155.6 billion by 2030. Bioethanol as a biofuel is the leading cause of this advance (Polaris 2022). Moreover, by 2050, waste generation is estimated to increase by up to 50%, which could substantially increase bioethanol production – with production forecasts of up to 140 billion liters in 2022 alone (Singh et al. 2022b).

Bioethanol can be used alone or added to other fuels such as gasoline. It is also used in the generation of energy by thermal combustion, in cells of thermochemical reaction cogeneration systems, and is the raw material for the chemical industry. Compared with gasoline, cellulosic ethanol reduces $CO_2$ emissions by up to 90%. In addition, the remaining residues from bioethanol production can be used as fertilizer in agricultural activity, closing the initial raw material cycle (Singh et al. 2022b). It can be produced from different biomasses rich in sugars, such as sugarcane bagasse, corn straw, cob, rice straw, wheat straw, cotton stalk, coconut husk, and pineapple leaf, in addition to countless other materials from agriculture (Boro et al. 2022); Singh et al. 2022b; Polaris 2022).

The agriculture sector generates another waste that stands out due to the high concentrations of carbohydrates, which are leftovers from the production and trade of fruits. Waste amounts to 1.3 billion tons per year, corresponding to a third of the world's food production, which could increase as the global population increases (Singh et al. 2022b). These residues can serve as a sustainable resource. Due to the high content of pentoses and hexoses in their composition, readily fermentable carbohydrates, they allow the use of environmentally friendly alternatives, such as the use of seawater, biomass composed of a mixture of fruit residues, or even remove high-value products such as D-Limonene and produce bioethanol in sequence (Bonatto et al. 2021; Zanivan et al. 2022; Singh et al. 2022b)

In addition to leveraging the production of bioethanol, the use of leftover fruit has other advantages, such as helping to meet the growing energy demand, avoiding inappropriate disposal or disposal in landfills, solving an environmental problem, and its low cost since the removal of this industrial waste needs to be carried out regardless of its destination (Singh et al. 2022b).

Depending on their origin, fruit residues contain polysaccharides such as cellulose and hemicellulose, and may also contain starch, proteins, and lipids. In addition to bioethanol, these compounds may be helpful in the production of other biofuels, such as butanol, biogas, biohydrogen, and biodiesel (Singh et al. 2022b).

#### 6.4.1.2 Biodiesel

Biodiesel has recently emerged as a vital substitute for petroleum diesel due to its renewability, lower GHG emissions, and low cost. It can be used pure or added in percentage to other fuels, the biofuel with the highest growth in several developed countries (Topare et al. 2022). Comparing biodiesel to diesel oil, $CO_2$ emissions reached a reduction of 78%. In addition, it has other advantages, such as it is obtained from different raw materials, especially non-edible oils and animal fat; it has lower toxicity, lower rate of emission of particulate matter, carbon monoxide, and aromatic hydrocarbons; it has a higher flash point; it is an excellent lubricant; and it is the fuel intended for diesel engines (Topare et al. 2022).

Biodiesel contributes to the circular economy as it can be produced from locally sourced materials. To reduce dependence on agricultural products, it is feasible to use available waste to generate a community-based biofuel through partnerships between companies in the community (Topare et al. 2022). This process contributes to a more sustainable, economically and environmentally advantageous approach. It is possible to produce biodiesel from different raw materials, such as edible and non-edible oils, such as rapeseed, sunflower, palm, castor bean, soy, mustard seed, peanuts, and canola, in addition to cooking oil residues, animal fats such as lard, tallow, and poultry fat. However, there are disadvantages to using edible oil, such as food shortages and deforestation for food crop production. Thus, an alternative is reusing oils and greases from industrial food processing or using mixed fats. Furthermore, microalgal lipids have recently emerged as favorable raw materials for biodiesel production (Kalita et al. 2022; Topare et al. 2022).

The predominant production route is transesterification, a reaction between oil and alcohol in the presence of a catalyst. Among the alcohols, the most used is methanol because it has a lower cost. Still, ethanol, amyl alcohol, and butanol, among others, can be used, and the oil:alcohol ratio is the determining parameter of this reaction. Generally, the chemical reaction is catalyzed by sulfuric acid, sodium hydroxide, or potassium hydroxide, which are considered homogeneous catalysts that provide higher yields in a shorter reaction time (Nayab et al. 2022; Topare et al. 2022). In the transesterification reaction, in addition to biodiesel, glycerol is also produced, which has high commercial value. However, using chemical catalysts impairs glycerol recovery, hampering the removal of used acids and bases, increasing the interference of free fatty acids and water, and increasing saponification reactions (Kalita et al. 2022). These problems can be overcome by replacing chemical catalysts with enzymatic ones. Enzymatic biocatalysts generally comprise lipases and enzymes from microorganisms such as fungi and bacteria. The use of enzymes as catalysts is more advantageous due to their selectivity, mild reaction conditions with less energy demand, and products with higher yield and better quality. In addition, they reduce the alcohol:oil ratio, and the enzymes can be reused in several cycles without losing their activity and minimizing the generation of effluents (Kalita et al. 2022).

Enzymatic catalysis is attractive and environmentally sustainable. However, the high cost of enzymes is an obstacle in the industrial production of biodiesel. There are some challenges in the viability of enzymatic catalyzes, such as using residues with greater availability to reduce costs, improve conversion, and increase the shelf life of lipases. As a positive point, when using enzymes, the lower generation of contaminants is highlighted, favoring the generation of cleaner products from this process (Kalita et al. 2022). In this way, the search for cheap raw materials becomes essential.

The main byproduct of the biodiesel production process is glycerol, a colorless, viscous, versatile liquid widely used as a raw material in the pharmaceutical, food, and cosmetics industries. However, depending on the origin of the oil used in biodiesel production, the glycerol may vary in its purity content, contain traces of methanol and water, and present a dark color. These characteristics are not desirable for commercialization, which implies the need for purification to be sold, and increased operations and costs (Haron et al. 2018). One possibility that has shown promise is the fermentation of glycerol to biohydrogen. This biotransformation is cheaper than purification, reduces the amount of waste, and is a clean process. Some studies report excellent biohydrogen production when using crude glycerol from biodiesel production (Haron et al. 2018).

The bioconversion of glycerol into biohydrogen stands out due to the diversity of routes that can be used, such as biophotolysis, photofermentation, dark fermentation, or even hybrid processes, which integrate two reaction systems. There are still some challenges to be overcome. Still, it is a good option that provides clean and harmless energy to the environment and stands out as a sustainable alternative to the fossil matrix (Haron et al. 2018).

#### 6.4.1.3 Biohydrogen

Another product that can also be obtained from the reuse of agricultural residues is biohydrogen. It is a fuel that does not have GHG emissions, has a high calorific value, is efficient in energy conversion, and has high density, being considered the ecological fuel of the future due to its green characteristics (Boro et al. 2022). Biohydrogen can be obtained through different processes from the activity of microorganisms that degrade substrates based on carbohydrates present in the most diverse residues, such as, for example, forestry and agricultural biomass, animal residues, food residues, sludge, industrial effluents, residues, urban solids, and wastewater, among others.

The best-known processes are bio-assisted photolysis of water by algae and cyanobacteria, photodecomposition of organic compounds by photosynthetic microorganisms, fermentation of organic waste, or even combining more than one process. Biohydrogen production via the biological route is a good and promising strategy that drives the circular economy and sustainable economic development (Cheng et al. 2022).

Biohydrogen production is interesting in the context of biorefinery, such as in the sugarcane industry, in which, after sugar processing, bagasse is used to produce nanocellulose, lignin, and biohydrogen. Comparing the stages of the processes of generation of these products separately, it has been found that they all have a more significant impact than the integrated form in the biorefinery system. In this context, the production of biohydrogen annexed to an already established process is highly advantageous, strengthens the principles of green chemistry, and has economic advantages (Katakojwala & Mohan 2022).

#### 6.4.1.4 Biobutanol

Biobutanol is a vital substitute for gasoline, has a very similar chemical structure, and requires little or no modification of vehicle engines, with a higher energy potential than gasoline and less volatility (Boro et al. 2022). Although its obtainment still predominates in the petrochemical route, it is possible to obtain it through biotechnological routes. It is a second-generation fuel, produced from the fermentation of biomass from different residues, not competing with human food and reducing dependence on first-generation fuels (Gwalwanshi et al. 2022). Among its qualities, resistance to friction, less corrosiveness, and flammability are important. It has a low vaporization enthalpy and consequently less volatility due to its high carbon content, greater thermal efficiency, and safety in transport. In addition, it is widely used as a solvent and chemical stabilizer in several areas of industry. It is a precursor in chemical synthesis processes (Gwalwanshi et al. 2022).

The biotechnological way of obtaining it is old and emerged in the First World War, the second-largest fermentation industry until the 1980s, through the acetone–butanol–ethanol (ABE) fermentation process. Currently, this process aims to replace the petrochemical matrix with lignocellulosic biomass, where the carbon source can be a wide range of raw materials, such as glucose-rich residues, effluents, food scraps, starch-rich biomass, wheat straw, rice, and corn, among others. The reaction medium occurs with the inoculation of fermenting microorganisms under mild process conditions, with relatively low temperatures (Liu et al. 2021). Biobutanol production via ABE is shown as an alternative to producing biofuels from solid waste management. It can be integrated into other routes aiming at greater productivity and less interference from the inhibitors generated by the processes, facilitating the separation of the compounds generated in the fermentation. In addition, it can use mixed cultures in consolidated bioprocesses, co-producing butanol, ethanol, acetone, biogas, and biohydrogen (Liu et al. 2021).

### 6.4.2 Enzymes

The sector of enzymes, known industrially as biocatalysts, has grown significantly over the last decade due to the growing global demand for cleaner and more sustainable technologies. In contrast to chemical catalysts, biocatalysts have high specificity. This characteristic confers excellent viability in industrial processes since a lower amount of secondary products are formed in the process, representing economic and environmental benefits. In addition to specificity, biocatalysts also work under mild pH and temperature conditions. As a result, industries have become interested in using enzymes due to their high catalytic activity compared with conventional catalysts and for acting with high efficiency under simple reaction conditions (Marques et al. 2017; BCC 2021).

The global market for enzymes in industrial applications is estimated to grow at an annual rate of 6.3%, reaching around US$8.7 billion by 2026 (BCC 2021). However, a critical analysis of the economics of an enzyme production plant reveals that almost 50% of the production cost is associated with investment capital and that the cost of raw materials accounts for approximately one-third. In recent years, studies in the literature have been focused on the search for increasingly efficient and optimized methods, as well as on improving the existing techniques until then of enzyme production economically and sustainably (Ravindran et al. 2018; Chavan et al. 2022). Agricultural, industrial, and urban wastes, mostly bio-based, are compounds that can be converted into value-added products, resulting in a greater return on investment applied in the global enzyme market (Chavan et al. 2022).

Due to their potential for thermal and pH stability, α-amylases have a wide range of industrial applications, making them better known and studied than the other enzymes in the group sustainably (Ravindran et al. 2018; Balakrishnan et al. 2021). This group of enzymes generates low-molecular-weight products, such as glucose and maltose, by breaking glycosidic bonds. Mainly because of their fundamental role in producing biofuels, cellulases are of great commercial importance. This group of enzymes break cellulose's glycosidic bonds, converting them into reducing sugars (Singh et al. 2021). In addition, proteases hydrolyze the peptide bonds that link amino acids together in polypeptide chains. Representing about 60% of the industrial enzyme market, they are used in various sectors, such as the food, pharmaceuticals, and leather industries sustainably (Ravindran et al. 2018). In addition to this sector, they are widely applied in making beer, bread, and paper, among others.

Due to the influence on the industrial market, combined with the high cost of producing these enzymes, usually using fungi and bacteria, the development of cheaper methods, such as waste, has gained prominence in this sector. Organic wastes such as soybean meal, corn pericarp, chicken feather, swine hair, banana peel, orange peel, rice bran, pineapple peel, and oil cakes, among many others, were applied for the production of enzymes (Preczeski et al. 2020; Almanaa et al. 2020; Balachandran et al. 2021; Balakrishnan et al. 2021).

In short, most of the residues listed in the literature are lignocellulosic, making them excellent raw materials for producing value-added products. However, the composition of these residues can affect the performance of biological processes, especially the yield of the products. Furthermore, the selection of the pretreatment method is substantial. Different pretreatment methods have been applied in recent decades to analyze their effect on the production of bioproducts. When chemical processes are used to treat corn pericarp residues for the production of α-amylase and protease, higher yields can be observed, with an increase of up to 34.3 and 1.7 times in enzymatic activities, respectively (Salim et al. 2017). Ultrasonic pretreatments of rice husk and olive pomace can improve the accessibility of fungi to the hemicellulose fraction of these residues, inducing the production of xylanases and cellulases, with an increase of up to 3 and 1.2 times in enzymatic activities, respectively (Leite et al. 2016).

### 6.4.3 Chemicals

After years of weak growth related to COVID-19 and trade tensions, the chemical industry appears to be on the mend. In 2021, chemical trade increased 26.8% yearly to $281.4 billion. Projections indicate that this sector should return to a growth trend by 2026. However, commodity demand, associated with market growth, should maintain high prices throughout the year (ACC 2022). As the global energy industry continues to shift from fossil fuels to renewables, coupled with the commitment of several countries to carbon neutrality by 2050, as a way of contributing to achieving the COP21 climate resolution, the sector is expected to focus on the search for new technologies to improve resource and energy efficiency in the production of chemicals and materials, thus boosting the decarbonization of the world energy matrix and sustainability.

Therefore, the use of waste for the production of valuable products is gaining strength as bio-based products reach a price on the market with the additional advantage of simultaneous reduction of waste generated, also contributing to the decarbonization of the world energy matrix while creating new byproducts as part of an increasingly circular economy. Agricultural residues and crops such as sugarcane bagasse, corn straw/stalks, cotton stalks, fruit residues, and cassava are used as substrates for producing lactic acid – a platform chemical for use in a wide range of industries. Studies were carried out to analyze the influence of corn stalk, cotton stalk, and fruit and vegetable residues (banana, orange, watermelon, pineapple, onion, potato, and tomato) on the production of succinic acid by *Actinobacillus succinogenes.* The results indicate that after hydrolysis of these residues, glucose concentrations increased significantly, resulting in yields of 17.8, 15.5, and 27.03 g/L of succinic acid, for corn stalk, cotton stalk, and fruit and vegetable residues, respectively (Li et al. 2010; Dessie et al. 2018).

Lignocellulosic residues used in biorefineries to produce value-added products are converted into soluble intermediate sugars via depolymerization reaction by enzymatic or acid hydrolysis. Alternatively, the sugars can be further derivatized into organic chemicals, such as 5-hydroxymethylfurfural, furfural, levulinic acid, and 2,5-furanedicarboxylic acid, for example (Ashokkumar et al. 2022).

### 6.4.4 Biopolymers

Environmental concerns have driven the need for biodegradable and/or bio-based polymers. The market has been growing considerably, with an annual growth rate of around 30% compared with conventional polymers. Polyhydroxyalkanoates (PHAs) are produced by several microorganisms that produce intracellular biopolymers, as long as there is carbon limitation in the culture medium. In addition, they are rapidly degradable and have low rigidity and high breaking strength, making them an excellent alternative to conventional polymers. However, due to the carbon source, the process costs are high for its commercialization (Tarrahi et al. 2020).

Traditionally, the carbon sources for the production of PHAs are mainly composed of pure carbohydrates (such as maltose, glucose, and starch), fatty acids and their derivatives, methanol, and alkanes. Residues such as watermelon, apple, coffee grounds/oil, whey, corn husks, and fish have been reported for the production of bioplastics (Koller et al. 2013; Araújo et al. 2018; Saratale et al. 2020; Liu et al. 2020). However, these materials cannot be used directly, requiring the pretreatment step for the production of the biopolymer. Notably, the production of PHAs depends on many factors, such as the quality of the waste, the microorganisms, the nutrients used, and the technologies for pre-processing (Saratale et al. 2020; Tarrahi et al. 2020). Thus, studies have been carried out in search of technologies that use renewable and low-cost substrates to produce PHAs.

### 6.4.5 Bioherbicides and Biofertilizers

Weed control has always been one of the significant issues in the agricultural sector. Herbicides have been widely used as a weed control strategy and, despite increasing productivity, their intensive use has been questioned, as it causes considerable environmental impacts. In this context, new products capable of replacing or reducing the use of conventional herbicides were developed. Bioherbicides are a significant group of control agents based on biologically active organisms or their metabolites. In addition, they are biodegradable, resulting in a shorter exposure time and risk of environmental contamination (Chilakamarry et al. 2022).

Agro-industrial waste, such as bananas, sugarcane, soybeans, coffee, rice husks/straws/bran, wheat bran, broom sorghum, sawdust, beet pulp, soybean meal, peanut pods, ferns, and cow manure, are used in the literature as a substrate for the production of bioherbicides with an efficiency greater than 80%. To select a suitable substrate for the bioherbicide formulation process, one must consider its availability, cost, and the quantity and quality of the microbial populations produced. This is because the production of viable people is influenced by the substrate (Naeimi et al. 2020; Chilakamarry et al. 2022).

However, using some organic residues, such as animal manure, can limit the use of products obtained through bioconversion due to the contaminants present in the substrate, which can pose risks of entry of toxic substances into the food chain and consequent health effects. An alternative to this problem is the use of insect larvae to convert these residues into value-added products. Larvae can significantly reduce, in a short period, the amount and content of nutrients present in animal manure, and this is related to the larvae's development time and the manure's constitution. This way, the residual substrate from waste bioconversion can be used as a biofertilizer (Matos et al. 2021).

In addition to animal manure, solid residues used after ethanol production can also be used as biofertilizers since they still contain a certain amount of nutrients after the process. Potato residues and pineapple leaves were evaluated for their potential to be used as fertilizer. The residues obtained after bioethanol were inoculated with different microorganisms to enrich nitrogen, phosphorus, and potassium. This process provided improved nitrogen, phosphorus, and potassium ratio, indicating the potential of using these residues as biofertilizers – which not only enriches the soil but also reduces the application of chemicals and promotes the concept of a circular economy (Chintagunta et al. 2016, 2017.

## 6.5 SUMMARY AND CONCLUSION

Based on the linear economy, the current economic model is unsustainable and unstable. Raw materials are extracted and transformed into discarded products right after use. Large-scale waste production will not cease but will increase and diversify due to agro-industrial activities and population growth, causing significant environmental damage since much of this waste is not adequately managed.

Because of this, the perspectives that move around the circular economy are systemic approaches aimed at benefiting, in addition to industries, society as a whole, the environment, and sustainable economic development, which is favored by the exponential growth of the urban fabric, which consequently requires an increase in the demand for inputs and chemical products, increasing the use of fossil residues and the production of organic residues. In this scenario, the biological conversion of waste into value-added products becomes a potential solution in terms of waste management since the use of these processes aims to generate minimal to zero waste, making it a sustainable and economical method carried out for making use of easily accessible raw material and tends to the production of enzymes, chemicals, biofertilizers, among others.

However, it is clear that the dynamics of innovation and market opening are crucial for the success and integration of this evolutionary process. It is equally necessary to highlight that until then, the sector is emerging in the global economy without a defined industrial structure. Currently, the planning of production resources is more significant in the biofuels process than in the manufacture of bioproducts, although it has a higher estimate of annual growth, standing above 20%. Therefore, the use of waste for the production of bioproducts and bioenergy should be explored even more to value agro-industrial and food waste that is generated in high quantity. Hence, by using waste from these sectors, in the future, the structuring of industries will be able to rely on the integrated production of biofuels and bioproducts, making the application of the circular economy an alternative to the current linear economy through environmentally friendly production processes.

## REFERENCES

ACC (American Chemistry Council). 2022. Mid-Year Outlook: A Good Year for US Chemistry Despite Risks to Growth. https://www.americanchemistry.com/chemistry-in-america/news-trends/press-release/2022/mid-year-outlook-a-good-year-for-us-chemistry-despite-risks-to-growth#:~:text=Basic%20chemicals%20will%20grow%20by,by%20strong%20demand%20and%20restocking (acessed August 7, 2022).

Almanaa, T. N., P. Vijayaraghavan, N. S. Alharbi, S. Kadaikunnan, J. M. Khaled, and S. A. Alyahya. 2020. Solid state fermentation of amylase production from *Bacillus subtilis* D19 using agro-residues. *Journal of King Saud University — Science* 32, no. 2, 1555–1561.

Araújo, C. S., A. M. C. Rodrigues, M. R. S. P. Joele, E. A. F. Araújo, and L. F. H. Lourenço. 2018. Optmizing process parameters to obtain a bioplastic using proteins from fish byproducts through the response surface methodology. *Food Packaging and Shelf Life* 16, 2–-30.

Ashokkumar, V., R. Venkatkarthick, S. Jayashree, et al. 2022. Recent advances in lignocellulosic biomass for biofuels and value-added bioproducts - A critical review. *Bioresource Technology* 344, 126195.

Astolfi, V., A. L. Astolfi, M. A. Mazutti, et al. 2019. Cellulolytic enzyme production from agricultural residues for biofuel purpose on circular economy approach. *Bioprocess and Biosystems Engineering* 42, 677–685.

Balachandran, C., A. Vishali, N. A. Nagendran, K. Baskar, A. Hashem, and E. F. Abd_Allah. 2021. Optimization of protease production from *Bacillus halodurans* under solid state fermentation using agrowastes. *Saudi Journal of Biological Sciences* 28, no. 8, 4263–4269.

Balakrishnan, M., G. Jeevarathinam, S. K. S. Kumar, I. Muniraj, and S. Uthandi. 2021. Optimization and scale-up of α-amylase production by Aspergillus oryzae using solid-state fermentation of edible oil cakes. *BMC Biotechnology* 21, 33.

BCC. 2021. Global Markets for Enzymes in Industrial Applications. https://f.hubspotusercontent00.net/hubfs/308401/BIO%20Report%20Overviews/BIO030L_Report_Overview.pdf?utm_campaign=BIO030L&utm_medium=email&_hsmi=141296193&_hsenc=p2ANqtz--LLFk2TMWRVtcQOMufqm-241TpiHRr5YO7LND_IGnl9FduAmZAvdapPxwYqbkHkc80KfAJRTgXG5aWUpmOnRcxj52Sliu3IbzFnKIsbcWO-6OqCOE&utm_content=141296193&utm_source=hs_automation (acessed July 31, 2022).

Bharathiraja, S., J. Suriya, M. Krishnan, P. Manivasagan, and S. K. Kim. 2017. Production of enzymes from agricultural wastes and their potential industrial applications. In: *Advances in Food and Nutrition* Research, eds. S. K. Kim, F. Toldrá, 125–148. Elsevier, Amsterdam.

Bonatto, C., T. Scapini, J. Zanivan, et al. 2021. Utilization of seawater and wastewater from shrimp production in the fermentation of papaya residues to ethanol. *Bioresource Technology* 321, 124501.

Boro, M., A. K. Verma, D. Chettri, V. K. Yata, and A. K. Verma. 2022. Strategies involved in biofuel production from agro-based lignocellulose biomass. *Environmental Technology & Innovation* 28, 102679.

Cajnko, M. M., J. Oblak, M. Grilc, and B. Likozar. 2021. Enzymatic bioconversion process of lignin: mechanisms, reactions and kinetics. *Bioresource Technology* 340, 125655.

Călin, M., D. Constantinescu-Aruxandei, E. Alexandrescu, et al. 2017. Degradation of keratin substrates by keratinolytic fungi. *Electronic Journal of Biotechnology* 28, 101–112.

Camargo, J. Z., V. M. Nascimento, I. Stefanello, et al. 2018. biochemical evaluation, molecular characterization and identification of novel yeast strains isolated from Brazilian savannah fruits, chicken litter and a sugar and alcohol mill with biotechnological potential for biofuel and food industries. *Biocatalysis and Agricultural Biotechnology* 16, 390–399.

Chavan, S., B. Yadav, A. Atmakuri, R. D. Tyagi, J. W. C. Wong, and P. Drogui. 2022. Bioconversion of organic wastes into value-added products: A review. *Bioresource Technology* 344, 126398.

Cheng, D., H. H. Ngo, W. Guo, et al. 2022. Enhanced photo-fermentative biohydrogen production from biowastes: An overview. *Bioresource Technology* 357, 127341.

Chilakamarry, C. R., A. M. M. Sakinah, A. W. Zularisam, R. Sirohi, I. A. Khilji, N. Ahmad, and A. Pandey. 2022. Advances in solid-state fermentation for bioconversion of agricultural wastes to value-added products: opportunities and challenges. *Bioresource Technology* 343, 126065.

Chintagunta, A. D., S. Jacob, and R. Banerjee. 2016. Integrated bioethanol and biomanure production from potato waste. *Waste Management* 49, 320–325.

Chintagunta, A. D., S. Ray, and R. Banerjee. 2017. An integrated bioprocess for bioethanol and biomanure production from pineapple leaf waste. *Journal of Cleaner Production* 165, 1508–1516.

Dedeke, G. A., F. O. Owagboriaye, A. O. Adebambo, and K. O. Ademolu. 2016. Earthworm metallothionein production as biomarker of heavy metal pollution in abattoir soil. *Applied Soil Ecology* 104, 42–47.

Dessie, W., W. Zhang, F. Xin, et al. 2018. Succinic acid production from fruit and vegetable wastes hydrolyzed by on-site enzyme mixtures through solid state fermentation. *Bioresource Technology* 247, 1177–1180.

FAO (Food and Agriculture Organization of the United Nations). 2022. Production Indices. https://www.fao.org/faostat/en/#data/QI (acessed June 28, 2022).

Gonzalo, G. D., D. I. Colpa, M. H. M. Habib, and M. W. Fraaije. 2016. Bacterial enzymes involved in lignin degradation. *Journal of Biotechnology* 236, 110–119.

Gupta, R., and P. Ramnani. 2006. Microbial keratinases and their prospective applications: an overview. *Applied Microbiology and Biotechnology* 70, 21–33.

Gwalwanshi, M., R. Kumar, and M. K. Chauhan. 2022. A review on butanol properties, production and its application in internal combustion engines. *Materials Today*: *Proceedings* 62, 6573–6577.

Haron, R., R. Mat, T. A. T. Abdullah, and R. A. Rahman. 2018. Overview on utilization of biodiesel by-product for biohydrogen production. *Journal of Cleaner Production* 172, 314–324.

Jain, A., S. Sarsaiya, M. K. Awasthi, et al. 2022. Bioenergy and bio-products from bio-waste and its associated modern circular economy: current research trends, challenges, and future outlooks. *Fuel* 307, 121859.

Jamaldheen, S. B., M. B. Kurade, B. Basak, C. G. Yoo, K. K. Oh, B. H. Jeon, and T. H. Kim. 2022. A review on physico-chemical delignification as a pretreatment of lignocellulosic biomass for enhanced bioconversion. *Bioresource Technology* 346, 126591.

Kalita, P., B. Basumatary, P. Saikia, B. Das, and S. Basumatary. 2022. Biodiesel as renewable biofuel produced via enzyme-based catalyzed transesterification. *Energy Nexus* 6, 100087.

Katakojwala, R., and S. V. Mohan. 2022. Multi-product biorefinery with sugarcane bagasse: process development for nanocellulose, lignin and biohydrogen production and lifecycle analysis. *Chemical Engineering Journal* 446, 137233.

Kaur, L., and S. Shah. 2022. Production of bacterial cellulose by *Acetobacter tropicalis* isolated from decaying apple waste. *Asian Journal of Chemistry* 34, 453–458.

Kim, A. L., S. Park, Y. K. Hong, J. H. Shin, and S. H. Joo. 2021. Isolation and characterization of beneficial bacteria from food process wastes. *Microorganisms* 9, 1156.

Koller, M., D. Sandholzer, A. Salerno, G. Braunegg, and M. Narodoslawsky. 2013. Biopolymer from industrial residues: life cycle assessment of poly(hydroxyalkanoates) from whey. *Resources, Conservation and Recycling* 73, 64–71.

Kumar, V., and S. K. Dwivedi. 2019. Hexavalent chromium reduction ability and bioremediation potential of Aspergillus *flavus* CR500 isolated from electroplating wastewater. *Chemosphere* 237, 124567.

Leite, P., J. M. Salgado, A. Venâncio, J. M. Domínguez, and I. Belo. 2016. Ultrasounds pretreatment of olive pomace to improve xylanase and cellulase production by solid-state fermentation. *Bioresource Technology* 214, 737–746.

Leonov, P. S., X. Flores-Alsina, K. V. Gernaey, and C. Sternberg. 2021. Microbial biofilms in biorefinery - Towards a sustainable production of low-value bulk chemicals and fuels. *Biotechnology Advances* 50, 107766.

Leschine, S. B. 1995. Cellulose degradation in anaerobic environments. *Annual Review of Microbiology* 49, 399–426.

Li, Q., M. Yang, D. Wang, et al. 2010. Efficient conversion of crop stalk wastes into succinic acid production by *Actinobacillus succinogenes*. *Bioresource Technology* 101, 3292–3294.

Liu, J., S. Fan, K. Bai, and Z. Xiao. 2021. Combining acetone-butanol-ethanol production and methyl orange decolorization in wastewater by fermentation with solid food waste as substrate. *Renewable Energy* 179, 2246–2255.

Liu, Y., S. Yang, and X. Jia. 2020. Construction of a 'nutrition supply-detoxification' coculture consortium for medium-chain-length polyhydroxyalkanoate production with a glucose-xylose mixture. *Journal of Industrial Microbiology and Biotechnology* 47, 343–354.

Marques, C. T., S. M. Golunski, B. Venturin, et al. 2017. Removal of a red dye using noncommercial peroxidase extracted from rice bran. *Environmental Quality Management* 27, 25–31.

Matos, J. S., A. T. M. S. Barberino, L. P. D. Araujo, I. P. Lôbo, and J. A. D. A. Neto. 2021. Potentials and limitations of the bioconversion of animal manure using fly larvae. *Waste and Biomass Valorization* 12, 3497–3520.

Meenakshisundaram, S, A. Fayeulle, E. Leonard, C. Ceballos, and A. Pauss. 2021. Fiber degradation and carbohydrate production by combined biological and chemical / physicochemical pretreatment methods of lignocellulosic biomass - a review. *Bioresource Technology* 331, 125053.

Muffler, K., M. Lakatos, C. Schlegel, D. Strieth, S. Kuhne and R. Ulber. 2014. Application of biofilm bioreactors in white biotechnology. In: *Productive Biofillms*, eds. K. Muffler, R. Ulber, 123–161. Springer, New York.

Naeimi, S., V. Khosravi, A. Varga, C. Vágvölgyi, and L. Kredics. 2020. Screening of organic substrates for solid-state fermentation, viability and bioefficacy of *Trichoderma harzianum* as12-2, a biocontrol strain against rice sheath blight disease. *Agronomy* 10, 1258.

Nair, L. G., K. Agrawal, and P. Verma. 2022. An overview of sustainable approaches for bioenergy production from agro-industrial wastes. *Energy Nexus* 6, 100086.

Nayab, R., M. Imran, M. Ramzan, et al. 2022. Sustainable biodiesel production via catalytic and non-catalytic transesterification of feedstock materials - A review. *Fuel* 328, 125254.

Ng, C. W. C., A. F. Ismail, M. M. Z. Makhtar, M. N. F. Jamaluddin, and H. A. Tajarudin. 2020. Conversion of food waste via two-stage fermentation to controllable chicken feed nutrients by local isolated microorganism. *International Journal of Recycling of Organic Waste in Agriculture* 9, 33–47.

Papadopoulos, M. C. 1986. The effect of enzymatic treatment on amino acid content and nitrogen characteristics of feather meal. *Animal Feed Science and Technology* 16, 151–156.

Pérez, J., J. Muñoz-Dorado, T. D. L. Rubia, and J. Martínez. 2002. Biodegradation and biological treatments of cellulose, hemicellulose and lignin: An overview. *International Microbiology* 5, 53–63.

Polaris, M. R. 2022. Bioethanol Market Share, Size, Trends, Industry Analysis Report, by Feedstock (Cereals & Starch, Wheat, Maize, Beet, Sugarcane, Others); by Industry; By Region; Segment Forecast, 2022–2030. https://www.polarismarketresearch.com/industry-analysis/bioethanol-market (acessed July 27, 2022).

Preczeski, K. P., C. Dalastra, F. F. Czapela, et al. 2020. *Fusarium oxysporum* and *Aspergillus* sp. As keratinase producers using swine hair from agroindustrial residues. *Frontiers in Bioengineering and Biotechnology* 8, 71.

Rahmani, A. M., P. Gahlot, K. Moustakas, A. A. Kazmi, C. S. P. Ojha, and V. K. Tyagi. 2022. Pretreatment methods to enhance solubilization and anaerobic biodegradability of lignocellulosic biomass (wheat straw): progress and challenges. *Fuel* 319, 123726.

Raina, D., V. Kumar, and S. Saran. 2022. A critical review on exploitation of agro-industrial biomass as substrates for the therapeutic microbial enzymes production and implemented protein purification techniques. *Chemosphere* 294, 133712.

Ramires, M. F., E. L. D. Souza, M. D. C. Vasconcelos, et al. 2020. Enzyme assays and toxicity of pig abattoir waste in *Eisenia andrei*. *Environmental Pollution* 260, 113928.

Ravindran, R., S. S. Hassan, G. A. Williams, and A. K. Jaiswal. 2018. A review on bioconversion of agro-industrial wastes to industrially important enzymes. *Bioengineering*, 5, 93.

Sadh, P. K., S. Duhan, and J. S. Duhan. 2018. Agro-industrial wastes and their utilization using solid state fermentation: a review. *Bioresources and Bioprocessing* 5, 1–15.

Salim, A. A., S. Grbavčić, N. Šekuljica, et al. 2017. Production of enzymes by a newly isolated *bacillus* sp. Tmf-1 in solid state fermentation on agricultural by-products: the evaluation of substrate pretreatment methods. *Bioresource Technology* 228, 193–200.

Saratale, G. D., R. Bhosale, S. Shobana, et al. 2020. A review on valorization of spent coffee grounds (SCG) towards biopolymers and biocatalysts production. *Bioresource Technology* 314, 123800.

Šelo, G., M. Planinić, M. Tišma, S. Tomas, D. K. Komlenić, and A. Bucić-Kojić. 2021. A comprehensive review on valorization of agro-food industrial residues by solid-state fermentation. *Foods* 10, 1–26.

Singh, A., R. R. Singhania, S. Soam, et al. 2022a. Production of bioethanol from food waste : status and perspectives. *Bioresource Technology* 360, 127651.

Singh, A., S. Bajar, A. Devi, and D. Pant. 2021. An overview on the recent developments in fungal cellulase production and their industrial applications. *Bioresource Technology Reports* 14, 100652.

Singh, S., A. Kumar, N. Sivakumar, and J. P. Verma. 2022b. Deconstruction of lignocellulosic biomass for bioethanol production : recent advances and future prospects. *Fuel* 327, 125109.

Sivaramakrishnan, R., B. Ramprakash, G. Ramadoss, S. Suresh, A. Pugazhendhi, and A. Incharoensakdi. 2021. High potential of Rhizopus treated rice bran waste for the nutrient-free anaerobic fermentative biohydrogen production. *Bioresource Technology* 319, 124193.

Sonawane, J. M., A. K. Rai, M. Sharma, M. Tripathi, and R. Prasad. 2022. Microbial biofilms: recent advances and progress in environmental bioremediation. *Science of the Total Environment* 824, 153843.

Srivastava, N., M. Srivastava, A. Alhazmi, T. Kausar, S. Haque, R. Singh, P. W. Ramteke, P. K. Mishra, M. Tuohy, M. Leitgeb and V. K. Gupta. 2021. Technological advances for improving fungal cellulase production from fruit wastes for bioenergy application: A review. *Environmental Pollution* 287, 117370.

Tan, F., J. Cheng, Y. Zhang, X. Jiang, and Y. Liu. 2022. Biotechnology for biofuels genomics analysis and degradation characteristics of lignin by *Streptomyces thermocarboxydus* strain DF3-3. *Biotechnology for Biofuels and Bioproducts* 15, 78.

Tan, L. S., M. S. M. Sueb, G. K. Chua, S. H. N. Shaarani, and S. Jamek. 2020. Screening of microorganisms from pineapple waste for fructooligosaccharides production. *IOP Conference Series: Materials Science and Engineering* 991, 012056.

Tarrahi, R., Z. Fathi, M. Ö. Seydibeyoğlu, E. Doustkhah, and A. Khataee. 2020. Polyhydroxyalkanoates (PHA): From production to nanoarchitecture. *International Journal of Biological Macromolecules* 146, 596–619.

Teigiserova, D. A., L. Hamelin, and M. Thomsen. 2020. Towards transparent valorization of food surplus, waste and loss: clarifying definitions, food waste hierarchy, and role in the circular economy. *Science of the Total Environment* 706, 136033.

Thurston, C. F. 1994. The structure and function of fungal laccases. *Microbiology* 140, 19–26.

Todhanakasem, T., A. Sangsutthiseree, K. Areerat, G. M. Young, and P. Thanonkeo. 2014. Biofilm production by *Zymomonas mobilis* enhances ethanol production and tolerance to toxic inhibitors from rice bran hydrolysate. *New Biotechnology* 31, 451–459.

Topare, N. S., R. I. Jogdand, H. P. Shinde, R. S. More, A. Khan, and A. M. Asiri. 2022. A short review on approach for biodiesel production: Feedstock's, properties, process parameters and environmental sustainability. *Materials Today: Proceedings* 57, 1605–1612.

Torre, I. D. L., M. Ravelo, S. Segarra, M. Tortajada, V. E. Santos, and M. Ladero. 2017. Study on the effects of several operational variables on the enzymatic batch saccharification of orange solid waste. *Bioresource Technology* 245, 906–915.

Treichel, H., G. Fongaro, T. Scapini, A. F. Camargo, F. S. Stefanski, and B. Venturin. 2020. *Utilising Biomass in Biotechnology: A Circular Approach Discussing the Pretreatment of Biomass, its Applications and Economic Considerations*. Springer, New York.

Wösten, H. A. B. 2019. Filamentous fungi for the production of enzymes, chemicals and materials. *Current Opinion in Biotechnology* 59, 65–70.

Xu, Q., Y. Liao, E. Cho, and J. H. Ko. 2020. Effects of biochar addition on the anaerobic digestion of carbohydrate-rich, protein-rich, and lipid-rich substrates. *Journal of the Air and Waste Management Association* 70, 455–467.

Yaashikaa, P. R., P. S. Kumar, and S. Varjani. 2022. Valorization of agro-industrial wastes for biorefinery process and circular bioeconomy : A critical review. *Bioresource Technology* 343, 126126.

Yamamura, S., Y. Morita, Q. Hasan, K. Yokoyama, and E. Tamiya. 2002. Keratin degradation: A cooperative action of two enzymes from Stenotrophomonas sp. *Biochemical and Biophysical Research Communications* 294, 1138–1143.

Zanivan, J., C. Bonatto, T. Scapini, et al. 2022. evaluation of bioethanol production from a mixed fruit waste by Wickerhamomyces sp. UFFS-CE-3.1.2. *Bioenergy Research* 15, 175–182.

Zhou, M., and X. Tian. 2022. Development of different pretreatments and related technologies for efficient biomass conversion of lignocellulose. *International Journal of Biological Macromolecules* 202, 256–268.

# 7 Biodrying of Municipal Solid Waste for Producing Refuse-Derived Fuel

*Chart Chiemchaisri, Wilai Chiemchaisri, Sakulrat Sutthiprapa, Ruchira Perera, Noppharit Sutthasil, Panida Payomthip, Komsilp Wangyao, and Sirintornthep Towprayoon*

## 7.1 INTRODUCTION

The municipal solid waste (MSW) generation in Asia has grown rapidly over several decades. Nevertheless, there are wide differences among countries in the region according to economic status, population density, urban lifestyle, food habits, geographical conditions, and other socio-economic and cultural factors (Asian Institute of Technology, 2004). Generally, MSW is defined as waste collected by municipalities or other local authorities and it typically includes household waste, commercial and institutional waste, and garden or park waste (IPCC, 2006). Recently, MSW generation rates in Asia were reported as 0.34–0.69 tons/cap/year, varying among different regions while ranging from 0.06 to 1.28 tons/cap/year for the reported countries (IPCC, 2019). Figure 7.1 shows MSW generation rates reported for different regions together with their high-moisture material, for example, food waste and garden waste, composition.

In Asia, MSW is mainly composed of food wastes with paper and plastics as other major components and they are commonly disposed of together in landfills or dumpsites (Idris et al., 2004). Due to its high food waste content, the commingled waste with limited source separation practice usually contains high moisture content and low calorific value (Shao et al., 2010). Modern waste

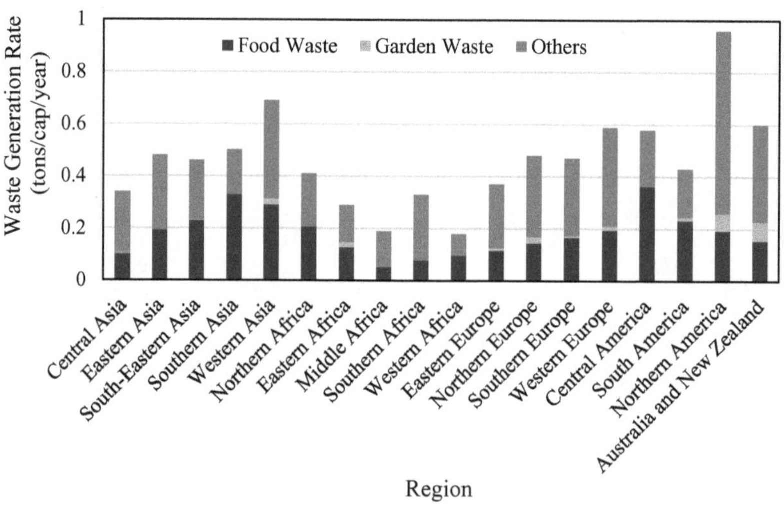


FIGURE 7.1 Waste generation rate in different regions.


DOI: 10.1201/9781003386902-9

management strategy focuses on the avoidance of solid waste disposal through recycling or transforming it into other useful products. One possible option is to convert MSW into refuse-derived fuel (RDF) for recovering energy from wastes. However, pretreatment of MSW especially aiming at moisture removal from MSW would be required to produce RDF, which meets the required quality for effective utilization in industries.

## 7.2 MECHANICAL BIOLOGICAL TREATMENT OF MUNICIPAL SOLID WASTE

Mechanical Biological Treatment (MBT) is a technique that is applied to treat or utilize an organic fraction of MSW. MBT is an integration of a mechanical stage and a biological stage to reduce the volume and stabilization of organic materials. Figure 7.2 presents typical treatment steps in the MBT process. First, the waste is subjected to a mechanical process aimed at sorting and/or pretreatment of waste. Subsequently, the segregated or pretreated waste proceeds to the biological treatment step for reducing the organic content of waste or preparing the residual waste known as RDF or Solid Recovered Fuel (SRF) for its utilization in the thermal process. In some applications, the biological treatment takes place before the mechanical treatment and the system may be called Biological Mechanical Treatment (BMT).

MBT of MSW is being widely implemented in several countries, especially in Europe. During the 2000s, there were more than 100 facilities in operation especially in Austria, Germany, and Italy with a total capacity of over 10 million tons of waste being processed annually with the larger plants capacitate at 600–1,300 tons/day (Diaz and Savage, 2006). Nevertheless, MBT application is still limited in Asia despite its potential being recognized in terms of the high organic fraction in MSW (Visvanathan et al., 2005).

An MBT plant can incorporate several different processes in a variety of combinations. It can be operated to achieve several different goals such as pretreatment of waste going to landfill, diversion of MSW from landfill through mechanical sorting for recycling, conversion into combustible biogas for energy recovery, and/or drying materials to use as energy recovery in terms of RDF (Park et al., 2013).

In the mechanical stage, the aim is to prepare materials for subsequent biological treatment by turning waste into smaller sizes and separating reusable or recyclable materials such as glass or metals out of the waste stream. The process may include size reduction/shedding of waste, separation of metals, size classification, density separation, heat/steam treatment, and screening and/or size reduction of outputs. The material left as residues can later be sold back into the market if it is processed adequately or it can be packed into balers to serve other purposes, such as recycling. In the biological stage, organic wastes are stabilized and inert materials could be used as RDF or

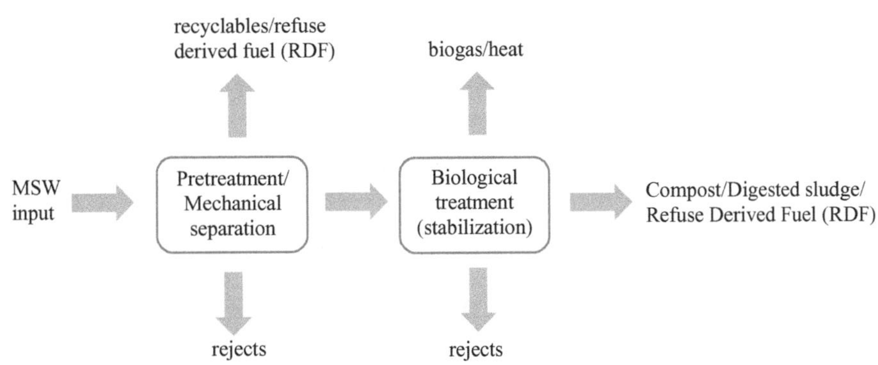


**FIGURE 7.2** Common MBT process.

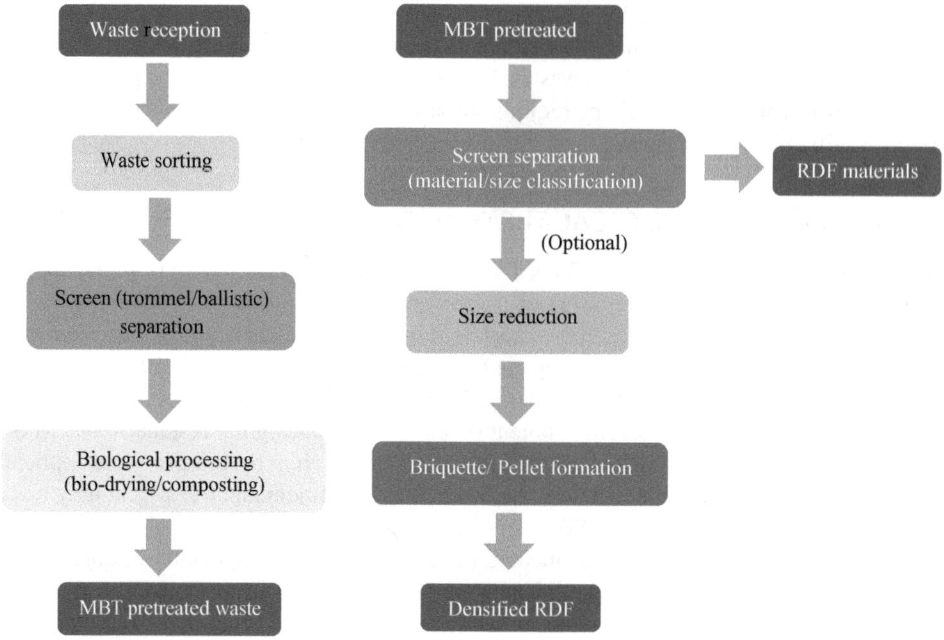


**FIGURE 7.3** Example of the process flow of MBT and RDF plants.

used in agriculture as a fertilizer and soil improvement. Biological treatment is usually performed in an enclosed system. The MBT-treated waste is recovered either through material/size separation or further processed into densified RDF through size reduction and briquette/pellet formation. A typical process flow in MBT and RDF plants is illustrated in Figure 7.3.

## 7.3 APPLICABILITY OF MECHANICAL BIOLOGICAL TREATMENT AND ITS POTENTIAL CHALLENGES

The MBT technology focuses mainly on the application of mechanical separation together with the biological treatment of organic components in MSW that cannot be diverted to recycling. It has various options depending on the aims of the final product, especially on the biological treatment component. In selecting suitable MBT schemes for local conditions, general criteria on technical, operational, and environmental aspects should be considered. Table 7.1 presents the evaluation of MBT technology options based on those criteria. Based on the comparison, MBT with continuous dry anaerobic digestion (dry AD) and MBT with biodrying are considered more suitable for treating moist MSW with low primary energy demand and low emission association.

The implementation of the dry AD process in developing countries is still limited. One of its technical problems is its lower and unstable energy recovery from the mixed MSW as compared with the thermal process. In the anaerobic process, only 5%–10% of the energy transferred from the complex organic wastes to methane is available for microorganisms for their growth compared with 50% for the equivalent aerobic process. Thus, anaerobic microorganisms tend to be slower growing and sensitive to changes in conditions. This leads to the fluctuation of methane content in the biogas product. On the other hand, biodrying was established to produce RDF as their primary output for further utilization in thermal processes. The majority of organic waste is converted to RDF regardless of its biodegradability, and the leftover small residuals are for final disposal. A major benefit of MBT with biodrying is the opportunity to incorporate all biogenic waste components of the input MSW, a carbon-neutral, alternative energy source, into a fuel product. The MBT with the

**TABLE 7.1**
**Comparison of MBT Technologies for Treating MSW**

| Criteria | MBT with Composting | MBT with Anaerobic Digestion | MBT with Biodrying |
|---|---|---|---|
| Proven/robust technology | +++ | ++ | +++ |
| Modular structure | ++ | ++ | ++ |
| Suitable for moist MSW | ++ | ++ | +++ |
| Minimization of landfill needs | + | + | +++ |
| Biogas product | - | ++ | - |
| Low primary energy demand | ++ | +++ | +++ |
| Utilization of energy in waste | + | +++ | +++ |
| RDF quality | + | ++ | ++ |
| Low emissions | ++ | +++ | +++ |
| Overall rating | Limited suitability | Good suitability | Good suitability |

*Source:* Modified from Kanning and Kettelsen (2013).
*+ fair, ++ good, +++ very good, - not applicable.*

biodrying process is, therefore, associated with low greenhouse gas emissions, which helped mitigate the global warming problem contributed by MSW management.

## 7.4 BIODRYING AS A PRETREATMENT OPTION FOR MUNICIPAL SOLID WASTE

Biodrying is considered an appropriate option for the pretreatment of MSW containing high moisture content in developing countries. Typically, the biodrying reactor receives pre-sorted MSW and produces a biodried output for the post-separation stage. Within the biodrying reactor, excess moisture is removed from the moist MSW, and its potential for energy recovery in the thermal process is enhanced. The evaporative heat is mainly derived from the aerobic decomposition of readily degradable organic matter that can take place in various types of reactors ranging from natural windrow piles to in-vessel reactors. These possible options made the process attractive for applications in economically developed and developing countries. Generally, biodrying reactors previously developed in developed countries use a combination of engineered physical and biochemical processes. Reactor design includes a container coupled with an aeration system; containers can be either enclosed or open tunnel halls, or rotating drums. In the biochemical process, aerobic biodegradation of readily decomposable organic waste occurs. On the physical effect, convective moisture removal is achieved through controlled, excessive aeration.

## 7.5 FUNDAMENTALS OF THE BIODRYING PROCESS

### 7.5.1 Principles of Biodrying

Biodrying is a drying technique that relies on the biological activities of microorganisms, bacteria, and fungi, to reduce the moisture content of wet biomaterial waste. The drying effect is achieved because of the microbial activity as well as forced aeration. As the microorganisms feed on the nutrients, carbon (C), nitrogen (N), and other elements available in the waste, heat is produced as part of the metabolic activities. This heat, assisted by air, is used to evaporate the excess moisture. In the biodrying process, the main drying mechanism is convective evaporation, using heat from the aerobic biodegradation of waste components and facilitated by mechanically supported airflow. The moisture content of the waste matrix is reduced through two main steps: (i) water molecules evaporate (i.e., change phase from liquid to gaseous) from the surface of waste fragments into the surrounding air and (ii) the evaporated water is transported through the matrix by the airflow and removed with the exhaust gasses. A limited amount of free water may seep through the waste matrix and be collected at the bottom of the biodrying reactor as leachate.

### 7.5.2 Difference Between Biodrying and Composting

Although the reactor configuration and physio-biochemical phenomenon are similar to composting, the operating condition in which it is controlled is different. Biodrying differs from composting in terms of the objectives of each process. Composting aims at fully decomposing biomass, removing odors, and killing pathogens. It produces a humus-like compost that can be used as a soil amendment. The duration of the natural compost pile can extend to several months. Composting is also used to stabilize the biodegradable organic material of MSW before landfill disposal, minimizing leachate and landfill gas formation. In contrast, the biodrying process aims to pretreat waste and remove as much moisture as possible in the shortest time while conserving organic matter to produce a high-quality RDF for energy recovery. The process duration varies with the waste material type and the system setup, but it typically should be within a few weeks for biodried piles and may require mixing at least once a week for moisture redistribution.

### 7.5.3 Benefits of Biodrying Application to MSW

Biodrying removes moisture from organic matter in MSW materials. The reduction of materials help to reduce the weight of those materials. It is also possible that the reduction in moisture results in a product that can be more easily handled and transported than in its original wet form. Additional benefits include the reduced potential for odors since the resulting material is well-aerated and partially stabilized. Finally, the resulting material with its lower moisture content is more suitable for energy production by a thermochemical process such as complete combustion, pyrolysis, or gasification. Mass balance determination in the biodrying process suggests that the majority of the waste mass of 60% could be recovered for the thermal process, whereas only a small fraction, for example, less than 15% of the original MSW, needs to be landfilled (Dziedzic et al., 2015).

### 7.5.4 Factors Affecting the Biodrying Process

Since the biodrying process is a combination of biological and physical processes, it is affected by several factors. The biological factors that directly influence the microorganisms include material composition, carbon-to-nitrogen (C/N) ratio, moisture content, and pH. Physical factors include pile geometry (shape), void ratio, airflow, and mixing intervals. An effective biodrying process can be achieved through appropriate reactor design and conditioning of various factors such as input material, microbial activities, and air supply, in combination with other process monitoring and control. The impacts of those influencing factors are described as follows:

- Input material

  The chemical and physical characteristics of the substrate are critical to the viability of the biodrying process. The particle size of waste materials and moisture content are some of the most important physical characteristics. The molecular size, complexity, and element composition of the chemical formula are included in the chemical characteristics. Increasing organic fraction of MSW in the feed materials promoted organic biodegradation and water removal during biodrying process provided that bulking agent was available (Zhang et al., 2018). Carbon-to-nitrogen ratio (C/N) is considered an important factor governing microbial activities in biodrying process, similar to that of composting process, with C/N of 20–30 reported to be optimal (Sadaka, 2011). Moreover, the complexity and molecule structure of substrate define its assimilability by microorganisms. The complex compounds are broken down by enzymes into intermediate compounds or into an element that microorganisms can utilize in their metabolism and synthesis of new cells. During biodrying of sewage sludge and kitchen waste, biodegradation of celluloses and hemicelluloses together with disappearances of tyrosine and tryptophan-like protein and formation of humic acid-like substances was observed (Zhang et al., 2020). Provision of the waste pre-hydrolytic stage a few days before high airflow supply during the biodrying stage could also improve moisture removals from the waste matrix (Somsai et al., 2015).

- Microbial activities

The microbial activities in the biodrying process derive the energy required for the evaporation of water from the waste matrix. The required heat is theoretically equal to the latent heat for vaporization, or enthalpy of vaporization which is provided mainly from aerobic biodegradation. In contrast to conventional drying which employs external sources of heat, aerobic decomposition of organic matter by microorganisms is an exothermic biochemical transformation that can rapidly raise waste matrix temperatures to the thermophilic range. Similar to composting process, maximum temperatures of 50–60°C for small-scale systems or up to 70°C for larger reactors could be reached. This heat is usually considered sufficient for drying, despite some losses from convection, radiation, and sensible heating of both the outlet air and discharged leachate. A small part of the external energy is also needed for aeration to convert to heat flow through the waste matrix usually driven by mechanical aeration.

A large variety of microorganisms can be found at different stages of biodrying. During the start-up under mesophilic temperature range, vegetative bacteria, bacterial endospores and mold fungi as well as ammonifiers could be detected at higher numbers but mainly bacterial endospores could proliferate to a higher number under thermophilic temperature condition of biodrying. Meanwhile, other microbial population including vegetative bacteria and other pathogens decreased due to unfavorable condition in the biodrying process such as high temperature and low moisture content (Malinowski and Wolny-Koładka, 2017).

- Air supply

The air supply is one of the main factors that affect to biodrying process. In the biodrying process, the main drying mechanism is convective evaporation, using heat from the aerobic biodegradation of waste components and facilitated by mechanically supported airflow. Therefore, sufficient air supply should be provided to facilitate water removal from the waste matrix. On the other hand, the provision of excessive aeration may reduce the temperature of the biodrying process (Colomer-Mendoza et al., 2013) thus yielding moisture removal through physical advection rather than heat derived from a biochemical process. Optimal ventilation should be provided to make biodrying more operative and rapid but also save energy and operation costs (Ab Jalil et al., 2016). Sadaka (2011) reported appropriate aeration rate between 0.2 and 0.6 $m^3$/min per 1 ton of waste for biodrying process. In addition to the effect on air supply rate, the mode of ventilation to facilitate airflow through the waste matrix was also found to affect the biodrying performance with continuous negative ventilation being the most preferable mode (Shao et al., 2012). The supply of hot-air, e.g., 65°C, during biodrying process of high-moisture (kitchen food) waste also led to better water removal and shorter biodrying period of 5 days (Li et al., 2022).

- Reactor design

  The effect of reactor design on biodrying efficacy was reported on the reactor height. The biodrying efficiencies in the reactor were reduced with moisture reduction reduced from 24.26% to 15.98% when waste matrix height was increased from 1.65 and 2.0 m during 10 days of operation (Tom et al., 2016a). However, it was also reported that an industrial-scale biodrying reactor with a waste layer height of 2.7 m could be operated when equipped with active aeration using supply and exhaust ventilator work continuously (Latosińska et al., 2022). A recent investigation was also performed using external natural heat (solar) input to improve the drying performance of mixed MSW (Ngamket et al., 2021).
- Process monitoring and control

  Process control can be performed by adjusting operational variables to be suitable for the biodrying condition with support from the monitoring of process parameters such as pH, moisture and temperature, etc. It is reported that pH of 5.5–8.0 and moisture content of 60%–65% are the appropriate conditions for biodrying (Sadaka, 2011). Due to the heterogeneity of the waste matrix especially in large-scale biodrying operations, the application of engineering process control in biodrying is quite challenging. The main difficulty is two different roles of treating waste, being mass to be dried as well as serving as the substrate to support microbial activities which generate the heat necessary for biodrying. High heterogeneity among the incoming and residual waste also made it difficult to determine the optimum condition for process control and the termination point of the biodrying process. However, the heterogeneity of waste composition also leads to high porosity and permeability of MSW which made it easily aerated (He et al., 2013).

## 7.6 APPROPRIATE CONDITIONS FOR BIODRYING

Physical and mechanical properties, such as moisture content, air-filled porosity, permeability, mechanical strength, and compaction of the matrix, have the potential to affect the resistance to the flow of air and, in turn, the level of airflow rate necessary for effective biodrying. Some of these properties could be beneficially conditioned by pre-processing the biodrying input to the reactor. Currently, the pre-processing strategy in most biodrying MBT plants is limited to coarse shredding, for example, at 150–300 mm maximum particle size. In biodrying, the moisture content can be reduced from 35%–55% to 10%–20% (Thomé-Kozmiensky, 2002). During aerobic biodegradation, around 0.5–0.6 g of metabolic water is produced per gram of VS decomposed. However, water losses during biodrying are much higher than the produced metabolic water. Water losses can be estimated using values of airflow rate and inlet–outlet air conditions and determined based on mass balance. The water balance should consider both metabolic water gains and evaporation–convection losses. Rada et al. (2007) reported overall weight losses of 25% as typical, but they could reach as high as 50% in small-scale operations (Zawadska et al., 2010). Various optimal temperatures have been reported for effective biodrying ranging from a lower value of 45°C to an upper value of 65°C (Velis et al., 2009).

Table 7.2 presents the summary of biodrying systems used at different scales as well as their operating conditions and achieved performance.

## 7.7 BIODRYING EXPERIMENT OF ASIAN MSW: AN EXPERIMENTAL STUDY

To demonstrate the applicability of the biodrying process to Asian MSW, laboratory-scale lysimeters of 0.30 m diameter and 1.50 m height (Figure 7.4a) were examined under different operating conditions to achieve appropriate biodrying performance. The lysimeters were equipped with moisture and temperature sensors and they were placed on a weighing machine. In the first experiment,

**TABLE 7.2**
**Application of Biodrying Systems with Their Operating Conditions and Performances**

| Reference | Biodrying System | Feed Waste | Operating Condition | Performance |
|---|---|---|---|---|
| Zhang et al. (2008) | Laboratory column reactor with thermal insulation (0.4 m diameter, 1.2 m height) | Raw MSW (32 kg, 210 kg/$m^3$ bulk density) | 0.45 $m^3$/kg/$m^2$.h aeration, waste turning every 2 days<br>Hydrolytic stage (0–4 days) with aeration of 10 min on/230 min off followed by aerobic stage (5–16 days) with aeration of 7 min on/23 min off. | • Combined hydrolytic and aerobic stage provided best biodrying performance and maximum moisture reduction<br>• No positive effect of inoculum addition on biodrying performance of combined hydrolytic and aerobic processes |
| Shao et al. (2010) | Laboratory columns with thermal insulation (0.4 m diameter, 1.2 m height) | Raw MSW (32 kg, 73% moisture content) | 0.056 $m^3$/kg/h intermittent aeration (7 min on/23 min off) waste turning every 2 days<br>16 days operation | • Treated waste with particle sizes of 45–60 mm and >60 mm had their calorific value drastically increased by 1.8 and 3.1 times and qualitied for European RDF standard<br>• Treated waste with particle sizes <45 mm had calorific value increased by 50%–60% but did not qualify for RDF standard<br>• Final moisture content of 48.3% and increase of calorific value of 157% was achieved |
| Shao et al. (2012) | Insulated columns with thermal insulation (0.4 m diameter, 1.2 m height) | MSW with glass and metal removal (32 kg, 280 kg/$m^3$ bulk density, 74% moisture content) | 0.014 $m^3$/kg/h with intermittent negative (7 min on/23 min off), continuous negative and intermittent positive ventilation modes (7 min on/23 min off) modes<br>16 days operation | • Intermittent positive ventilation mode provided greater microbial activity yielding highest water removals<br>• However, continuous negative ventilation had highest water loss to volatile solid loss suggesting its highest biodrying efficiency<br>• Final waste contained 31.3%–32.8% moisture content |

*(Continued)*

**TABLE 7.2 (*Continued*)**
**Application of Biodrying Systems with Their Operating Conditions and Performances**

| Reference | Biodrying System | Feed Waste | Operating Condition | Performance |
|---|---|---|---|---|
| Yuan et al. (2018) | Laboratory-scale column reactor (0.36 m diameter, 0.6 m height) | MSW (30 kg, 695 kg/m$^3$ bulk density, 71.47% moisture content) | 0.2–0.6 L/kg dry wt./min under continuous aeration mode<br>0.35, 0.6 L/kg dry wt./min under intermittent aeration mode, waste<br>turning frequencies at 0, once every 2, 3 and 6 days<br>18 days operation | • A continuous aeration rate of 0.5 L/kg dry wt./min and turn every 3 days gave the best biodrying performance<br>• Changing the aeration pattern and turning frequency did not affect biodrying performance when total aeration volume was constant |
| Tom et al. (2016b) | Pilot-scale insulated chamber (0.6 m diameter, 2.0 m height) | Mixed MSW (109 kg, 20 cm particle size limit, 61.25% moisture content, 200 kg/m$^3$ bulk density) | 0.528 m$^3$/kg/day aeration supplied,<br>33 days operation | • Max temperature of 50°C–50°C reached after 4 days<br>• Wet weight of wastes gradually reduced and became constant after 20 days<br>• Final moisture content of 48.5% or 20.81% moisture reduction was achieved |
| Tambone et al. (2011) | Full-scale plant with biodrying pile of 5.5 m height, total treatment capacity of 49,485 tons/year | MSW (6,600–7,000 kg of waste mass in biodrying pile, 32.72% moisture) | Air suction provided at the bottom of the pile to control exit gas temperature of 50°C.<br>The aeration was provided over 14 days for moisture removal | • Average air flow rate of 20.3 m$^3$/ton dry wt./h yielded maximum temperature of 53°C of exit gas temperature.<br>• The biodried waste contained calorific value 41% higher than that of untreated waste<br>• Final moisture content was reduced from 327.2 kg/ ton waste to 112.8 ton in biodried waste or 214.4 kg/ton waste (65.5%) water removal |
| Latosińska et al. (2022) | Full-scale biodrying reactor (6.1 × 6.1 × 4.0 m, 150 m$^3$, 2.7 m waste height) | MSW (50–60 tons per batch, <80 mm particle size, 56.5% moisture) | Aeration provided continuously through supply ducts at the bottom and exhaust vent at the top of reactor<br>14 days operation period | • The biodried waste had its calorific value 45% higher than that of primary waste<br>• Final moisture content of 30%–35% was achieved in biodried waste, yielding 50% moisture decrease |

**FIGURE 7.4** Biodrying units applied at different scales in Thailand. (a) laboratory-scale lysimeters, (b) pilot-scale lysimeters, (c) full-scale MBT biodrying systems (windrow pile-left, bunker type-right)

the lysimeters were supplied with different aeration modes, that is, continuous aeration at 0.2, 0.5, 1, and 2.25 L/min and intermittent aeration of 0.4 L/min (1 h on 1 h off and 3 h on 3 h off). The lysimeters were loaded with 22.0 kg MSW collected by a municipality in Thailand corresponding to an initial waste density of 325 kg/m$^3$.

Under continuous aeration rate of 0.2 L/min, it was found that the temperature elevated and fluctuated in a daily pattern, whereas moisture content in wastes started decreasing significantly on the fifth day and remained relatively constant after 7 days. The development of biodrying in the lysimeter was promoted by the heat released from the aerobic decomposition of organic wastes that could be developed within 5 days. The moisture and temperature patterns observed during the lysimeter operation at an aeration rate of 0.5 L/min had similar patterns as those observed at the aeration rate of 0.2 L/min. When the lysimeters were operated at a higher aeration rate of 1 and 2.25 L/min, physical evaporation of wastes was promoted, evidenced by a sudden drop of moisture content in the wastes to 20%–40% during the first day and it gradually declined afterward. The experimental results suggested that the aeration rates of 0.2 and 0.5 L/min could promote biodrying conditions in the lysimeters and the optimum aeration rate was 0.2 L/min in this study.

Figure 7.5a shows the comparison of moisture content and heating value of wastes before and after the biodrying process at different aeration rates. The moisture content in MSW (weight basis) could be reduced from 65.2% to 11.9%–14.1% in the lysimeters. The aeration rate of 0.2 L/min yielded the lowest moisture content in treated waste while producing a biodried product with a sufficient calorific value of 4,630 kcal/kg. Meanwhile, the highest aeration rate of 2.25 L/min gave the highest remaining heating value as the moisture loss was mainly due to physical evaporation.

The analysis of gas samples collected from the lysimeter operated under continuous aeration of 0.2 L/min revealed the lowest $O_2$ (15%) and highest $CO_2$ concentrations, being 2%–4% initially and increased to 10%, as compared with the other lysimeters. $CH_4$ was found generally at low concentration (<10 ppm). Meanwhile, $N_2O$ was detected at lower concentration in most lysimeters among which slightly higher concentrations (up to 2.5 ppm) were detected in the lysimeters operated at higher aeration rates. The moisture reduction during biodrying was also affected by aeration mode with different results obtained during continuous and intermittent aeration application. While the lysimeter operated under continuous aeration had moisture content in wastes decreased drastically after 5 days, the moisture reduction started earlier after 3 days under intermittent mode with the same amount of air supply. Nevertheless, condensation of moist air was taken place during the aeration off period resulting in higher moisture content detected at the upper part of the waste layer.

As shown in Figure 7.5b, the moisture content in MSW could be reduced from 63.6% to 14.8%–37.7% in the lysimeters. The reduction of moisture was mainly taken place upon heat development in the lysimeters indicated by the increased temperature of MSW after 3–4 days. Among different aeration modes applied, continuous aeration of 0.2 L/min yielded the lowest average moisture content (18.2%–18.5%) and was consistent along the depth of the lysimeter. As a result, the calorific value of 4,573–4,845 kcal/kg was obtained. Meanwhile, intermittent aeration at the same amount of airflow yielded lower moisture content (14.8%–16.7%) at the lower part where aeration was supplied but the moisture content was found higher in the upper part due to condensation of moist vapor, and hence non-uniform drying was achieved when intermittent aeration was supplied.

Comparing different intermittent aeration patterns applied, intermittent aeration of 0.4 L/min (3 h on/3 h off) yielded the best results in terms of moisture removal. The biodrying was developed well at a higher aeration rate (0.4 L/min) over a longer period of interval (3 h). Nevertheless, the condensation also took place in the top part of the waste layer during the aeration off period resulting in the remaining moisture content within the waste matrix. The results also revealed high mass

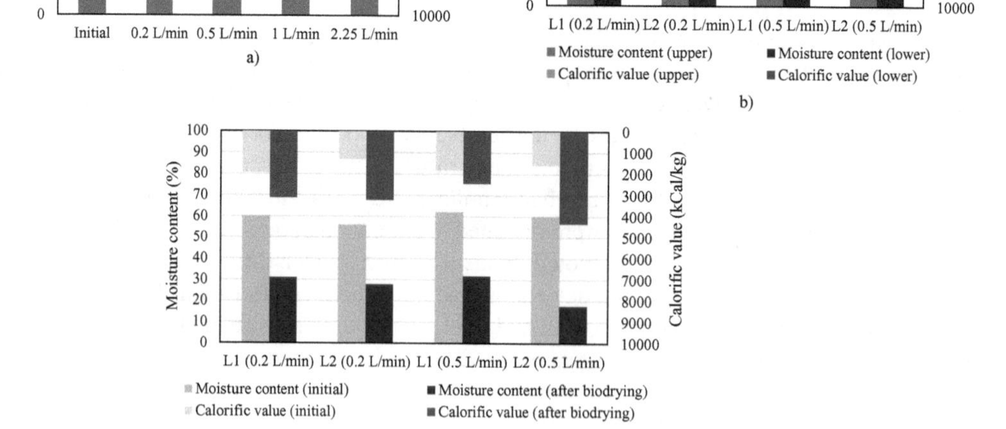


**FIGURE 7.5** Changes in moisture content and calorific value of wastes during biodrying under different conditions: (a) different aeration rates, (b) different aeration modes, and (c) different waste composition.

losses of solid wastes under intermittent aeration than those under continuous aeration. Therefore, the supply of aeration in continuous mode at a lower aeration rate was better than the intermittent air supply if the organic matter in MSW is to be preserved. To examine the effect of waste composition on biodrying performance, the lysimeters were supplied with continuous aeration of 0.2 and 0.5 L/min using MSW (26.0 kg) with different compositions, that is, L1 (50% food waste, 10% paper, 20% plastic, and 20% yard wastes) and L2 (40% food waste, 10% paper, 10% plastic, and 40% yard wastes). The lysimeters were operated for 14 days.

Figure 7.5c shows the biodrying performance in the lysimeters with different waste compositions at different aeration rates. The lysimeter with higher food waste contains higher moisture. At 0.2 L/min aeration rate, the moisture content of MSW containing more food waste (L1) could be reduced to only 30.5%–31.2% after 14 days, whereas they were 24.1%–31.3% in less food waste condition (L2). Inadequate moisture removals in both lysimeters could be due to ineffective airflow through the waste matrix for microbial activities to produce heat from microbial degradation. A higher aeration rate (0.5 L/min) yielded better results only in lysimeters with low food waste and high bulking (yard waste) conditions. Under such conditions, the moisture could be effectively reduced by about 70% and the drying was quite uniform throughout the lysimeter. The waste composition with lower food waste percentage and higher bulking conditions facilitated air flow through the waste matrix and promoted biodrying conditions.

## 7.8 BIODRYING SYSTEM AT LARGER (PILOT AND FIELD) SCALE OPERATIONS IN THAILAND

Pilot-scale lysimeters (0.5 m width and 2.0 m height) were set up and operated at the MBT plant of a cement industry in Thailand (Figure 7.4b). The lysimeter was equipped with an aeration pipe and perforated plate to facilitate airflow at the bottom. Negative aeration was performed by providing air suction rates of 0.5, 0.75, and 1.0 $m^3$/kg waste/day. During the biodrying period of 7 days, homogeneous temperature development was observed when the aeration rate of 0.5 $m^3$/kg waste/day was supplied. At this aeration rate, the highest weight and moisture reductions of 25% and 66% were achieved. The final product contained the lowest moisture content of 22% and the highest LHV of 4,938 kcal/kg (Payomthip et al., 2022).

A full-scale MBT biodrying system has also been implemented at the municipality level in Thailand since 2005. The process comprises an uncomplicated homogenization followed by biological treatment in windrow piles (Punin et al., 2014). This MBT process is located within a landfill site with a total area of approximately 368,000 $m^2$. The site receives MSW collected within the municipality at a daily rate of approximately 100 tons. At the facility, 4 MBT windrow piles were configured in different shapes depending on the site condition. The first pile was configured with 4 m height (Figure 7.4c left). The windrows are naturally ventilated, supported by pallets placed at the bottom of windrow piles. The biologically treated wastes further proceed to the separation steps consisting of manual separation and trommel screen to separate plastic wastes (> 40 mm), the organic fraction (10–40 mm), and soil-like materials (<10 mm). After the MBT processing of 1,500 tons of fresh MSW, about 60% of incoming wastes were reduced within 9 months. Simultaneously, about 75% of organic wastes were decomposed (Sutthasil et al., 2020). Further separation of treated waste yielded 13.6% of soil-like materials, 7.3% of organic fraction, and 79.1% of plastic wastes, which were subsequently used in RDF production (Punin et al., 2014). The plastic waste and organic fractions had their moisture content and calorific value qualified for RDF European standards (European Commission, 2003).

Another full-scale MBT biodrying system operated in Thailand is bunker type (Figure 7.4c right). The biodrying unit employed a waste channel equipped with a vertical screw mixer. The system received about 20 tons of MSW daily. Aeration could be promoted through the use of a screw mixer (4 sets), which operated periodically along the waste channel (4 m width x 40 m length x 2 m waste height). The biodrying period lasted about 7–10 days in this unit. During biodrying, moisture content in MSW could be reduced from about 50% to 20%–30% and the biodried waste could be utilized as RDF in nearby cement factories.

## 7.9 SUMMARY AND CONCLUSIONS

Biodrying technology is an attractive option for the treatment of MSW containing high moisture content. It allows the use of a major fraction of organic wastes presented in mixed MSW as a substitute fuel in RDF form for energy recovery in industrial processes. Despite its fundamental simplicity, the design and operation of the biodrying process need to be optimized for the local MSW and environmental conditions. The technology has been examined at various scales ranging from laboratory to full-scale implementation in Thailand using different system configurations and operating conditions. Nevertheless, universal design and operation criteria for optimization of biodrying reactors applied to Asian MSWs are still needed to be further developed.

## ACKNOWLEDGMENT

This work was carried out as an activity under the Strengthening Network Program on enabling waste management schemes in response to disaster and climate change supported by the Office of Higher Education Science Research and Innovation Policy Council of Thailand.

## REFERENCES

Ab Jalil, N.A., H. Basri, N.E. Ahmad Basri, and M.F.M. Abushammala. 2016. Biodrying of municipal solid waste under different ventilation periods. *Environmental Engineering Research* 21(2): 145–151.

Asian Institute of Technology. 2004. *Municipal Solid Waste Management in Asia*. Asian Regional Research Program on Environmental Technology (ARRPET), Pathumthani, Thailand.

Colomer-Mendoza, F.J., Prats, L.H., Martinez, F.R., Izquierdo, A.G., and Guzman, A.B.P. 2013. Effect of airflow on biodrying of gardening wastes in reactors. *Journal of Environmental Sciences* 25(5): 865–872.

Diaz, L.F., and G.M. Savage. 2006. Production and quality of refuse derived fuel (RDF). In: Proceedings of Biomass and Waste to Energy Symposium, Venice, Italy.

Dziedzic, K., B. Łapczyńska-Kordon, M. Malinowski, M. Niemiec, and J. Sikora. 2015. Impact of aerobic stabilisation and biodrying process of municipal solid waste on minimisation of waste deposited in landfills. *Chemical and Process Engineering* 36(4): 381–394.

European Commission-Directorate General Environment. 2003. Refuse Derived Fuel, Current Practice and Perspectives: Quality Standards for Solid Recovered Fuel. European Commission.

He, P., L. Zhao, W. Zheng, D. Wu, and L. Shao. 2013. Energy balance of a biodrying process of organic wastes of high moisture content: A review. *Drying Technology* 31: 132–145.

Idris, A., B. Inanc, and M.N. Hassan. 2004. Overview of waste disposal and landfills/dumps in Asian countries. *Journal of Material Cycles and Waste Management* 6: 104–110.

IPCC. 2006. 2006 *IPCC Guidelines for National Greenhouse Gas Inventories,* Vol.5: *Waste.* Intergovernmental Panel on Climate Change, Institute for Global Environmental Strategies, Japan.

IPCC. 2019. 2019 *Refinement to the 2006 IPCC Guidelines for National Greenhouse Gas Inventories*, Vol. 5: *Waste*. Intergovernmental Panel on Climate Change, Switzerland.

Kanning, K., and K. Ketelsen. 2013. *MBT: Best Technology for Treatment of Moist MSW AD and/or Biodrying Prior to Energy Recovery*. In: Proceedings of the Waste to Resources 2013, Cuvillier Verlag, Göttingen.

Latosińska, J., M. Żygadlo, and M. Dębicka. 2022. The biological drying of municipal waste in an industrial reactor-a case study. *Energies* 15: 1039.

Li, J., T. Ju, L. Lin, F. Meng, S. Han, Y. Meng, Y. Du, M. Song, T. Lan, and J. Jiang. 2022. Biodrying with the hot-air aeration system for kitchen food waste. *Journal of Environmental Management* 319: 115656.

Malinowski, M., and K. Wolny-Koładka. 2017. Microbiological and energetic assessment of the effects of the biodrying of fuel produced from waste. *Ecological Chemistry and Engineering S* 24(4): 551–564.

Ngamket, K., K. Wangyao, S. Patumsawad, P. Chaiwiwatworakul, and S. Towprayoon. 2021. Quality improvement of mixed MSW drying using a pilot-scale solar greenhouse gas biodrying system. *Journal of Material Cycles and Waste Management* 23: 436–448.

Park, J.R., U.H. Cho, B.R. Kim, Y.G. Kim, J.H. Nah, S.J. Bae, G.M. Leem, and D.H. Lee. 2013. Experimental case study of bio-drying mechanical-biological treatment in Korea. In: Proceedings of the Waste to Resources 2013. Cuvillier Verlag. Göttingen.

Payomthip, P., S. Towprayoon, C. Chiemchaisri, S. Patumsawad, and K. Wagyao. 2022. Optimization of aeration for accelerating municipal solid waste biodrying. *International Journal of Renewable Energy Development* 11(3): 878–888.

Punin, W., S. Maneewan, and C. Punlek. 2014. The feasibility of converting solid waste into refuse-derived fuel 5 via mechanical biological treatment process. *Journal of Material Cycles and Waste Management* 16: 753–762

Rada, E., A. Franzinelli, M. Taiss, M. Ragazzi, V. Panaitescu, T. Apostol. 2007. Lower heating value dynamics during municipal solid waste bio-drying. *Environmental Technology* 28(4): 463–469.

Sadaka, S., 2011. *Partial Composting for Biodrying Organic Materials.* Cooperative Extension Service, University of Arkansas, Fayetteville.

Shao, L., X. He, N. Yang, J. Fang, F. Lu, and P. He. 2012. Biodrying of municipal solid waste under different ventilation modes: Drying efficiency and aqueous pollution. *Waste Management and Research* 30: 1272–1280.

Shao, L., Z. Ma, H. Zhang, D. Zhang, and P. He. 2010. Bio-drying and size sorting of municipal solid waste with high water content for improving energy recovery. *Waste Management* 30: 1165–1170.

Somsai, K., T. Tondee, and S. Kerdsuwan. 2015. Effect of airflow on moisture removal of rotary biodrying reactors. *Jurnal Teknologi* 78(5–6): 57–63.

Sutthasil, N., C. Chiemchaisri, W. Chiemchaisri, T. Ishigaki, S. Ochiai, and M. Yamada. 2020. Greenhouse gas emission from windrow pile for mechanical biological treatment of municipal solid wastes in tropical climate. *Journal of Material Cycles and Waste Management* 22: 383–395.

Thomé-Kozmiensky, K.J. 2002. Processing concepts for substitute fuels. *Mineral Process* 43: 11–20.

Tom, A.P., A. Haridas, and R. Pawels. 2016a. Biodrying process efficiency: Significance of reactor matrix height. *Procedia Technology* 25: 130–137.

Tom, A.P., R. Pawels, and A. Haridas. 2016b. A sustainable technology for treatment of municipal solid waste with high moisture content. *Waste Management* 49: 64–72.

Tambone, F., Scaglia, B., Scotti, S., Adini, F. 2011. Effects of biodrying process on municipal solid waste properties. *Bioresource Technology* 102: 7443–7450.

Velis, C., P.J. Longhurst, G.H. Drew, R. Smith, and S.J.T. Pollard. 2009. Biodrying for mechanical-biological treatment of wastes: A review of process science and engineering. *Bioresource Technology* 100(11): 2747–2761.

Visvanathan, C., J. Tränkler, and C. Chiemchaisri. 2005. Mechanical biological pre-treatment of municipal solid waste in Asia. In: Proceedings of International Symposium MBT 2005, Hanover.

Yuan, J., D. Zhang, Y. Li, J. Li, W. Luo, H. Zhang, G. Wang, and G. Li. 2018. Effects of the aeration pattern, aeration rate, and turning frequency on municipal solid waste biodrying performance. *Journal of Environmental Management* 218: 416–424.

Zawadska, A., L. Krzystek, and S. Ledakowicz. 2010. Autothermal biodrying of municipal solid waste with high moisture content. *Chemical Papers* 64(2): 265–268.

Zhang, D., P. He, T. Jin, and L. Shao. 2008. Bio-drying of municipal solid waste with high water content by aeration procedures regulation and inoculation. *Bioresource Technology* 99: 8796–8802.

Zhang, D., W. Luo, J. Yuan, and G. Li. 2018. Co-biodrying of sewage sludge and organic fraction of municipal solid waste: Role of mixing proportions. *Waste Management* 77: 333–340.

Zhang, D., Z. Xu, G. Wang, N. Huda, G. Li, and W. Luo. 2020. Insights into characteristics of organic matter during co-biodrying of sewage sludge and kitchen waste under different aeration intensities. *Environmental Technology and Innovation* 20: 101117.

# Section C

## Case Studies

# 8 Large-Scale Rotary Drum Composting for Urban Organic Waste Management

*Suryateja Pottipati and Ajay S. Kalamdhad*

## 8.1 INTRODUCTION

Due to population growth and fast urbanization, organic solid waste generation is predicted to expand dramatically over the following years, posing a hazard to long-term development and depleting natural resources (Shah et al., 2022). Rapid urbanization, population growth, and rising living standards have exacerbated waste disposal issues (Gupta et al., 2021). Organic waste currently accounts for around 46% of all global solid waste. According to global waste management assessors, global organic waste production in 2016 reached 2.01 billion tonnes (Chavan et al., 2022a). Fruit and vegetable post-harvest waste reached up to 20% globally, trailing only root, tuber, and oil-producing crops at 25% (FAO, 2019). However, with exponential population growth and supply chain unsteadiness, a growing concern has resulted in more waste being generated globally (Ganesh et al., 2022). In India, fruit and vegetable production has steadily increased over the decades (from 88,977,000 MT/year in 2013–2014 to 97,358,000 MT/year in 2018–2019) (Srivastava and Balakrishnan, 2022).

Mixed waste poses significant issues to its producers. Complying with all relevant rules is sometimes difficult, though not impossible, and disposal sites, when accessible, may be costly (Lipton, 2004). Municipal solid waste deposited at an open dumpsite has more than 50% organic material, which creates a tremendous quantity of heat in the dumpsite subsurface owing to the long-term thermal decomposition of organic waste. Heat production from biological and chemical waste degradation results in spontaneous waste fires, which have serious environmental consequences (Chavan et al., 2019). Because of interactions with other substances placed in landfills, open burning of municipal solid waste (MSW) emits $CO_2$, CO, $SO_X$, $NO_X$, particulate matter, and hazardous gas emissions. Organic waste destined for landfills may be redirected to decentralized waste treatment facilities, reducing the risk associated with its disposal. The discharge of organic litter into the sea or any river water body may significantly impact the aquatic ecosystem and constitute a significant long-term threat to the marine ecosystem. Open dumping of biowaste in developing nations has severe environmental and social consequences, limiting governments' ability to fulfil sustainable development objectives (Chavan et al., 2022b).

Although the requirements for overall management can only be met if organic recycling is not neglected, information on the adaptation to biowaste composting of existing mechanical–biological treatment (MBT) plants designed initially to stabilize organic MSW is limited (Bernat et al., 2022). Existing composting methods require at least 2–3 months. Long residence times and low maturation efficacy have become major impediments to developing and disseminating organic wastes. Furthermore, specific composting systems, such as Windrow, operate in the open area, where the material and operational parameters dramatically change with the seasons and weather, making product quality challenging to manage (Wei et al., 2021). Rotary drum composting (RDC) can convert a variety of organic feedstock into nutrient-dense compost (Pottipati et al., 2022b).

DOI: 10.1201/9781003386902-11

In this chapter, a full assessment of the waste situation and its negative consequences is explored, as well as the sources of various organic wastes and their bioconversion possibility through adaptable large-scale reactor techniques. The primary goal of this chapter is to illustrate the possibility for two reactor-based waste management strategies to be adopted on a large scale, either alone or in combination, for the urban environment. The procedures presented in this chapter may help worldwide urban authorities' organic waste treatment sites develop a value-added end product from various organic wastes generated in communities.

## 8.2 INDIAN CITIES' APPROACH TOWARD WASTE MANAGEMENT

### 8.2.1 Open Dumping

More than 90% of generated MSW are unsubstantiated landfilling or uncontrolled dumping on the outskirts of towns and cities, which has profound environmental consequences in terms of greenhouse gases emission and numerous health concerns (Saha et al., 2010). Around 40% of garbage is disposed of in landfills globally, with less than 10% disposed of in appropriate and regulated systems such as sanitary landfills. The vast majority occurs in open dumps, common among most low-to middle-income nations (Muchangos and Tokai, 2020). The waste management issue of uncontrolled dumping and open burning has been highlighted in global venues such as the current United Nations Sustainable Development Goals for public health and natural resources. Furthermore, local governments in poor and middle-income nations are increasingly coping with the consequences of open dumping, driving a change in policy and action in favour of developing sustainable options for ultimate MSW disposal (Muchangos and Tokai, 2020; Rodić and Wilson, 2017).

Given the proximity to rivers, the climate and geological circumstances of the region, and the possible long-term pollution from produced leachate and landfill gas, open dumping concerns human health, security, and the environment (Wilson et al., 2016). It also contributes to global warming by emitting greenhouse gases (GHG) and other atmospheric pollutants (Srigirisetty et al., 2017). Commonly associated with open dumping is garbage open burning, which emits GHGs and significantly impacts public health. According to estimates, the open burning of garbage causes 270,000 premature deaths worldwide yearly, adding to around 5% of global GHG emissions (Alexander Cogut, 2016; Muchangos and Tokai, 2020).

Boragaon dumpsite in Guwahati city of India is one example where the dumpsite was initiated in an eco-sensitive area and very near lakes (Dipor Bil) and river catchments. According to the Central Pollution Control Board (CPCB) report in 2017, wet waste is 53.69% of total waste generated in Guwahati city, the primary disposal site of Guwahati is the Boragaon dumpsite (Figure 8.1). Leachate from dumpsites includes massive volumes of organic matter (biodegradable but resistant to biodegradation), with humic-type elements including ammonia-nitrogen, heavy metals, and chlorinated organic and inorganic salts. The difficulties caused by leachate are minor in the early phases of a dump site, but the problems become pretty concerning as the dump site matures. Because the leachate's characteristics might alter the receiving body, it could be a groundwater table or a surface water table.

### 8.2.2 Thermal Treatment

Incineration is the most widely used thermal waste-to-energy conversion process (Gimžauskaitė et al., 2022). MSW incineration is being promoted as a potential solution to India's growing waste problem. Six such plants are currently operational, and many more are in various stages of planning, building, and tender documents under the National Plan of the "Clean India Mission" (Gupta et al., 2021). Solid and liquid wastes, both hazardous and non-hazardous, may be burned to produce ash. Incineration is the technique of thermally combusting solid or liquid waste. The kind of incinerator utilized is determined by the garbage that will be burned (El-Haggar, 2007). Both primary and

**FIGURE 8.1** Boragaon dumpsite of Guwahati.

secondary burning chambers are needed in practically all kinds of waste incinerators to complete the combustion and oxidation required to achieve the desired destruction and disposal efficiency (Figure 8.2). The main chamber's job is to volatilize the waste's organic portion. The secondary chamber has to heat the evaporated organics to the temperature required for complete oxidation (El-Haggar, 2007).

#### 8.2.2.1 Benefits of Incineration

- Almost all kinds of waste can be managed.
- Through air pollution control systems, incinerators are designed to prevent air pollution.
- The ash produced by burning accounts for just around 10% of the total amount of solid waste.
- Energy may be recovered through incineration in various methods, including gas-to-water heat exchange. Water is transformed into steam, and then used to create power through steam turbines.

#### 8.2.2.2 Disadvantages of Incineration

- Expensive system.
- Skilled staff requirement.
- Requirement of external energy.
- Not entirely environmentally friendly.
- Requirement of homogeneity in waste.

Air pollution control methods are too costly. On the other hand, the gases and ash produced by incineration are exceedingly hazardous. They generate air pollution if not adequately regulated, which may be hazardous to human health (El-Haggar, 2007). Incineration systems are designed

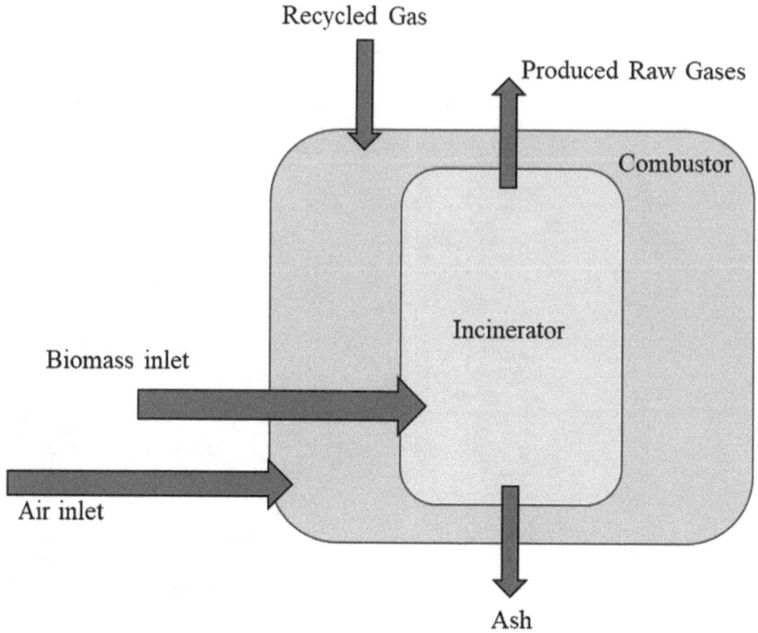


**FIGURE 8.2** Line diagram of the thermal degradation process.

to minimize MSW by providing controlled burning of garbage at high temperatures. However, incineration is typically seen as environmentally unfavourable since it has been observed that when 1 kg of solid waste is burned, 0.51 kg of carbon dioxide equivalent emissions is produced. In 2011, approximately 12% of the total MSW created (17.9% of MSW disposed of) was transformed by burning with energy recovery. 34.9 million tonnes of wasted food were disposed of in landfills or incinerated in 2011. A trash incineration project's success depends on accurate data on predicted waste amounts and characteristics (Trabold and Nair, 2018).

### 8.2.3 Mechanical–Biological Treatment

It is estimated that India generates approximately 141,064 tonnes of waste per day, of which 90% is collected, but only 27% is recycled (Pottipati et al., 2022a). MBT systems are often employed in managing leftover MSW and independently collected waste materials (Figure 8.3). The mechanical component of MBT plants is intended to maximize secondary material recovery, for example, plastics, metals, glass and so-called Refuse-Derived Fuel (high-calorific alternative fuel). The biological component of an MBT plant stabilizes the organic percentage of MSW, which is mechanically separated from residue MSW on a sieve (Bernat et al., 2022).

However, a study was carried out to examine the physicochemical properties, fertilizing potential, and heavy metal polluting prospects of MSW organic manure produced in 29 cities across the country. Except for a few samples, the results showed that all other samples had normal pH and EC. In general, the organic matter and primary nutrients Nitrogen and Phosphorus levels of MSW composts are lower than those of rural waste composts. Heavy metal levels in composts from larger cities were higher for Zn, Cu, Cd, Pb, Ni, and Cr. Regarding fertilising factors and heavy metal concentration, partial separation at the composting site did not considerably enhance compost quality. Most MSW composts failed to meet the quality control guidelines of the "The Fertilizer (Control) Order 1985" in total organic matter, total P, total K, and heavy metals Cu, Pb, and Cr. According to the definition, most of the compost samples did not fall into any of the classes and were therefore declared inappropriate for use. According to the legal restrictions of various nations, relatively few

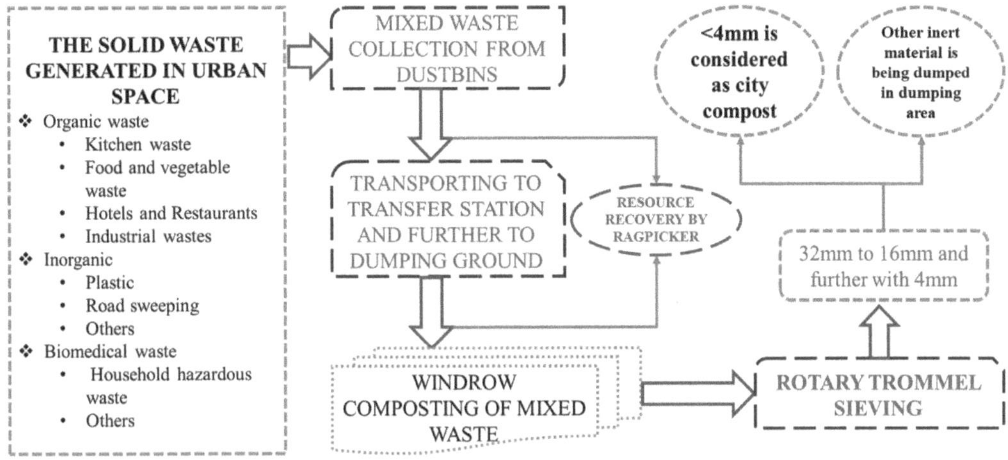


**FIGURE 8.3** Line diagram of the pyrolysis reaction.

compost samples (made from origin-separated biogenic wastes) were found to be in marketable categories, and some samples were determined to be appropriate only for limited usage (Saha et al., 2010). However, it is predicted that the treatment operations might produce considerable amounts of microplastics and their concentration in the organic output. Organic MBT outputs as a source of microplastics are yet unknown (Sholokhova et al., 2022).

## 8.3 SOURCES OF ORGANIC WASTE IN THE COMMUNITY

### 8.3.1 Residential

Individuals generate 0.5–1.0L of trash per day, with a minimum of 25% to a maximum of 35% organic matter and 10%–60% moisture content, depending on location (Owusu-Ansah et al., 2022). A complicated series of patterns influence household food waste, and disrupting these patterns may allow for a reduction in this massive quantity of food waste (Schuster et al., 2022). Socioeconomic characteristics, in particular, have been demonstrated to be a substantial predictor of recycling practices at the household level (Lo and Liu, 2018). The creation of home garbage varies in quantity depending on the size of the community, but it is not something that can be neglected as a substantial source of organic waste from the community, particularly in metropolitan regions.

### 8.3.2 Commercial

According to FAO (2019), 1.3 billion tonnes of food are lost or squandered annually, accounting for one-third of all food produced for human use (Lévesque et al., 2022). The Hotel, Restaurant, and Institutional segments significantly contribute to food waste generation in major cities. Further, India's function halls and festivals are also an unavoidable source of waste generation with potential organic portions. Horticultural waste in the community produces enormous amounts, which can be used as a bulking agent for composting.

### 8.3.3 Industrial Sludges

Food processing units, ayurvedic product–producing industries, organic sludges from dairy industries, and sometimes petroleum wastes (Paul Choudhury et al., 2021) can also serve as a possible source of organic waste from the community. Another potential source from urban areas is the sludge produced by the sewage treatment plants upon direct application to soil poses numerous environmental issues (Pottipati and Kalamdhad, 2022).

## 8.4 LIMITATIONS OF CENTRALIZED FACILITIES

Many experts have shown that relying on centralized MSW management design is inefficient based on the experiences of other nations. The rising MSW volumes necessitate the growth of garbage collection fleets and the elongation of transportation itineraries, which contribute to traffic congestion, local pollution, and road degradation. Increasing MSW volumes also encourages the construction of new big disposables and incinerators to replace those that are fast reaching capacity. These characteristics make centralized MSW management costly and unsustainable in reality, which may reduce city resilience and increase public health and environmental hazards. In the face of rising urbanization, the possibility of lowering massive disposal capabilities has been acknowledged as a significant global problem (Kuznetsova et al., 2019; Yadav et al., 2017). Figure 8.4 illustrates one such plant in Guwahati city of India.

### 8.4.1 Disadvantages of Centralized Facilities

- High installation cost.
- Huge land requirement.

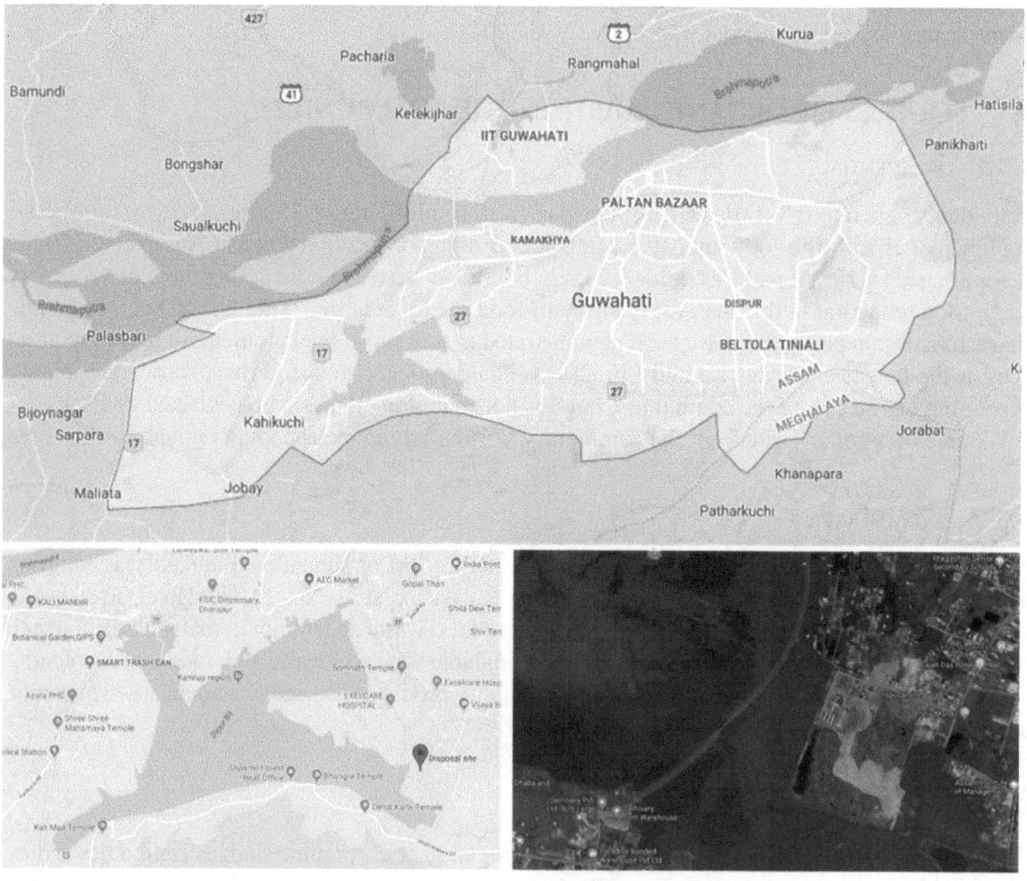


**FIGURE 8.4** Location of Boragaon dumpsite.

- Requirement of skilled labourers.
- Numerous environmental issues.
- Often involves high transportation costs due to the remote location of the plant.
- Poor quality end product because of the diverse nature of waste.

### 8.4.2 Case Study-Boragaon Dumpsite Guwahati

The 50-ton composting facility at the Boragaon dumpsite is an example of poor composting technology management. The plant has not been operational since 2012, although it was built in 2008. The large expenditures were only maintained for 4 years with an undesirable outlet product. The primary issue is inappropriate technology utilization; the facility is often centralized, and the feed used was a MSW. Composting is an organic decomposition process. However, if the input contains various components, the result will never be what is expected.

Studies on the Dipor Bil have been conducted around the Boragaon dump site, demonstrating a reduction in the pH value of water samples (that threaten acidic pH) around the dump site and is due to surface runoff from the dump site interacting with lake water. Moreover, the concentration of soluble oxygen, a primary source of oxygen for aquatic ecosystems, has dropped to less than 4 mg/L, indicating that if this situation persists for an extended period, it will significantly impact the lake ecology (Figure 8.5). Heavy metal concentrations have also been investigated in the Dipor

**FIGURE 8.5** Non-functional centralised facility at Boragaon dumpsite.

Bil zone near the dumpsite. The findings indicate that the water in Dipor Bil near the dumpsite has heavy metal concentrations that exceed the limits of drinking water regulations. This conclusion is concerning for the influence of heavy metals not only in the vicinity of the dumpsite but also throughout the Dipor Bil since the water level rises and falls owing to the passage of water from the Vashista and Brahmaputra rivers throughout the year (Dash et al., 2021).

## 8.5 IMPORTANCE OF DECENTRALIZED WASTE MANAGEMENT APPROACHES

Decentralized systems, on the other hand, operate on smaller yet controllable sizes. The idea allows garbage to be processed and stored, increasing its value. The relationship between high-value rubbish and waste dealers may be found in practically every nook and cranny of our cities. The current network of such dealers and informal employees forms an unauthorized knowledge pool that evaluates the value of garbage (Iyer, 2016). A study conducted for Bilaspur City in India determined that allocating a decentralized composting facility is more cost-effective than a centralized approach. Furthermore, a decentralized composting operation has lower transportation costs than a centralized one using the GIS platform (Rathore et al., 2022). Decentralized waste management options can also be appropriate when the location of a large-scale centralized plant is less appealing in terms of sustainability due to transportation impacts (González et al., 2020).

### 8.5.1 Advantages of Decentralized Facilities

- Onsite management of waste.
- Treatment feasibility according to the composition of the waste.
- Generates low-skilled jobs at the community level.
- Reduces the time required for overall treatment.
- Reduces the organic waste stress on landfills.
- Resource recovery can be effective.
- The community itself can utilize the produced compost.
- Reduces the impact on the environment.

## 8.6 OPTIMISED WASTE MANAGEMENT STRATEGIES

According to the research, aerobic biodegradation strategies are the most effective in tackling the problem (Blossey et al., 1994; Saha et al., 2017). For example, biodegradation utilizing an in-vessel rotational drum composter may be finished in 20 days and result in a stable final product (Jain and Kalamdhad, 2019; Kauser et al., 2020). Because the thermophilic phase in a rotating drum composter may be reached physiologically within 24 h (Kalamdhad et al., 2009a), these established thermophilic temperatures accelerate the substantial degradation of organic matter. These created thermophilic temperatures hasten the significant decomposition of organic matter. When compared to the first 7 days, degradation was much lower during this period (Hazarika and Khwairakpam, 2018; Jain and Kalamdhad, 2018; Kalamdhad et al., 2009b; Kauser et al., 2020; Sharma et al., 2018; Singh et al., 2012; Varma and Kalamdhad, 2016).

Because earthworms perform a biological function, vermicomposting can produce nutrient-rich compost. Earthworms expedite organic waste breakdown, boost carbon, nitrogen, and phosphorus circulation, and improve soil fertility (Devi and Khwairakpam, 2020). Vermicomposting adds to degradation by fine-grinding substrates, which increases porosity and microbial activity. This time-consuming method requires at least 45 days (Biruntha et al., 2020; Deepthi et al., 2021). Several studies have used vermicomposting to mature predegraded organic wastes and found that it reduced the degradation duration and improved product quality (Kalamdhad et al., 2009a; Varma et al., 2016; Varma and Kalamdhad, 2016). Vermicomposting removes heavy metals. Earthworms' heavy metal bioaccumulation has been examined to discover whether they may be utilized as bioindicators of soil pollution and to assess bioavailability and toxicity (Liu et al., 2005).

Experiments have combined in-vessel RDC with vermicomposting to shorten the time necessary for vermicomposting. The tests were successful on a limited scale and included different urban wastes, such as vegetable waste (Pottipati et al., 2021) and sewage sludge (Pottipati and Kalamdhad, 2022). Traditional biodegradation procedures take 45–120 days to degrade garbage, depending on the material and approach. The innovative process of combining both processes can generate outstanding grade vermicompost in 27 days, whereas existing biodegradation procedures cannot. Aerobic biodegradation of terrestrial weeds is an economically feasible biological control strategy that generates low-skilled employment at the community level. This technology may be an excellent alternative to pesticides used on terrestrial weeds with proper rotational drum composter maintenance and earthworm culture. Revenue generating might also be achievable if the experiments were conducted in pilot-scale reactors (Pottipati et al., 2022b).

### 8.6.1 Large-Scale Rotary Drum Composters

For the Indian waste scenario, only one study was conducted on a pilot-scale RDC in 2009. Long-term experiments were undertaken on a 3.5 $m^3$ demonstration plant under continuous flow circumstances to understand the compost dynamics of high-rate composting and the operational characteristics of a rotating drum composter (Kalamdhad et al., 2009a). The study aimed to look at how several physicochemical and biological parameters changed throughout 7 days of high-rate composting of vegetable waste, calf manure, dry tree leaves, and sawdust in a rotating drum composting system. Furthermore, this research aims to determine a suitable maturation process for primarily stabilized compost from a horizontal flow rotary drum composter (Figure 8.6). For a 20-day maturation period, two alternative composting processes, Windrow and vermicomposting, were investigated. The effects of seasonal fluctuation on maturation were also considered. Only one such study was conducted on real-time waste at the research level. The potential outcomes of the study demonstrate the feasibility of large-scale rotary drum composters in urban waste management.

#### 8.6.1.1 Prerequisites of the Rotary Drum Composting Process

a. Organic waste

   The first need is a continual stream of organic waste, which may be met by daily communal vegetable waste as well as various commercial and industrial sources. This entails collecting and preparing substrate at the treatment location on a daily basis.

b. Shredder and homogenizer

   A shredding machine and a homogeniser are required for the facility (to reduce manpower). Composting is more successful when the substrate is shredded to 1–2 cm prior to composting (Pottipati et al., 2022b). The waste is homogenised with the inoculum (typically

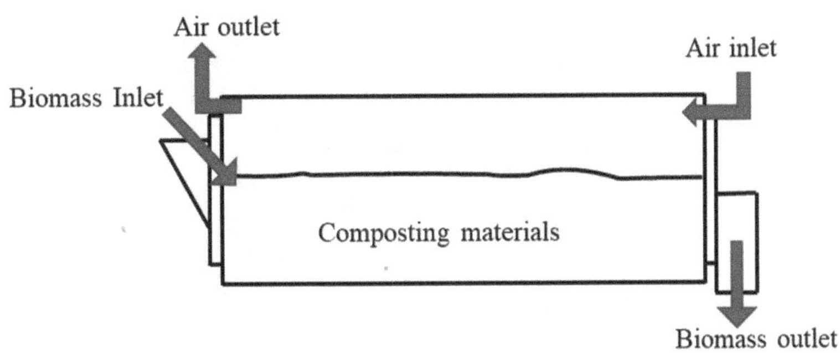


**FIGURE 8.6** Line diagram of large-scale rotary drum composter.

prepared compost) and the bulking agent using a homogeniser (horticultural waste). This shredding and homogenising operation will fulfil the feedstock preparation requirements for any composting process.

c. Reactor commissioning

Composting reactors are typically batch-type composters in which the inoculum is combined with the substrate and supplied to the reactor. The reactor itself may serve as a source of inoculum in large-scale composting employing in-vessel composting. The waste retention duration of 7–10 days assists the feedstock in mixing and homogenising the newly supplied waste with the existing waste within the reactor. Since the reactor's commissioning, the early days of feeding while the reactor is empty. The inoculum must be fed initially, followed by the first feed until the reactor is half-full. Furthermore, the microbial consortia that emerged from the reactor will serve as an inoculum for future feeds.

## 8.6.2 Large-Scale Vermicomposting Approach

### 8.6.2.1 Bag Vermicomposting

The common method of vermicomposting uses a plastic bag of 0.5–5 $m^3$ capacity plastic bags or concrete blocks. Usually used as a batch reactor for a minimum of 45–50 days and adopted in rural areas of India (Figure 8.7a). The material will initially be shredded and mixed using rice straw or dry leaves, and further cow manure will be used to fill the block. Waste and cow dung will be filled in layers, and then the earthworms will be subjected to the degradation purpose. Once the block is filled with waste material, it will be covered with a porous sheet to reduce sunlight penetration and also reduce the evaporation of moisture. This process also needs daily supervision. This technique is very less adopted in the urban scenario since it is both time-intensive and space-demanding.

**FIGURE 8.7** (a) Bag vermicomposting reactor, (b) Stack vermicomposter and (c) procedure to fill Stack vermicomposting units.

#### 8.6.2.2 Stack Vermicomposting

The novel method of vermicomposting is where we can produce a high quantity of product in very little space with little investment in the infrastructure. Numerous MSW management facilities in Guwahati are adopting this technique in managing their organic waste. The material will be shredded and mixed with the inoculum (prepared compost/cow dung if the first time) and then fed to one small reactor of 20–30 kg capacity, and these reactors will be stacked using iron frames (Figure 8.7b). Once the reactor completes the degradation time scale (45–60 days), the material will be then subjected to sieving to make the material fine for packing and then used for various purposes.

#### 8.6.2.3 Combination of Both Techniques

Considering the time required for the degradation individually and the compromising quality of only RDC, researchers are looking forward to further optimization. Studies conducted by Kalamdhad et al. (2009b) showed the use of in-vessel composting. If the feed is continuous, the duration of active composting of 7–10 days will be effective. Further, vermicomposting of the fresh material is said to take a minimum of 45 days. A study conducted by Pottipati et al. (2021) concluded that the total duration of composting can be optimized for vermicomposting on a batch scale to 27 days if both techniques are utilized only in the active phases. This means using RDC for the initial 7–10 days and then vermicomposting the partially degraded material using vermicomposting. The study also concluded that the duration of vermicomposting can be limited to 20 days post RDC, which develops further optimization of the process and produces vermicompost in 27 days. Figure 8.8 illustrates the layout of the process.

#### 8.6.2.4 Prerequisites of Vermicomposting Process

a. Vermiculture

   Vermicompost is biomass that has been decomposed by vermiculture. Vermiculture is provided in India via several local rural organisations as well as commercially. In bag vermicomposting, 1–2 kg of earthworms are given to 100–500 kg of organic waste, while Stack vermicomposting depends on the Stack size and the quantity of trash supplied to every reactor.

b. Horizontal sieve cum vermi separator

   When the vermicomposting process is finished, the finished product must be sieved using a 4 mm sieve before being packaged and distributed. There are several kinds of

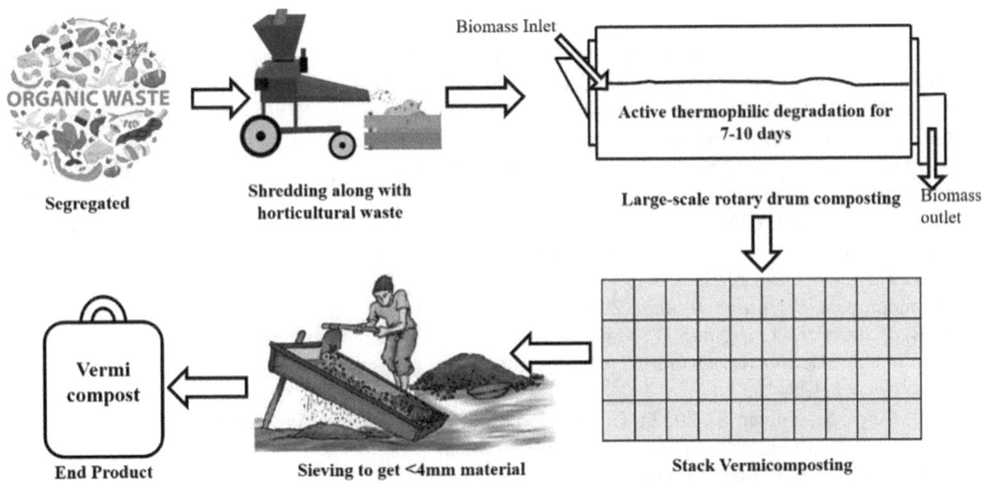


**FIGURE 8.8** Layout of the combined two-stage composting process.

sieves available in commercial marketplaces for this purpose. The horizontal shaking sieve rose to prominence due to the simplicity of its functioning mechanism.

## 8.7 PERSPECTIVE FOR THE TECHNOLOGY ADOPTION AS A DECENTRALIZED WASTE MANAGEMENT TOOL

The optimized composting technique discussed in the study has a huge scope as a tool in decentralized waste management techniques. Biodegradable waste has been an unavoidable part of MSW. Daily production can be seen as a potential source of organic waste. The disposal of organic waste, along with dry waste, is a major problem for techniques such a pyrolysis, incineration, and also producing refuse-derived fuels. The organic waste produced in urban spaces often ends up in landfills and is also one of the reasons for leachate generation and GHG from landfills. The technique discussed can reduce the community's stress from organic waste, and if implemented at the community level, the technique can serve to produce manure, which the community can use for a variety of purposes. The organic waste produced in urban areas can now find a new way of recycled as manure.

## 8.8 SUMMARY AND CONCLUSIONS

Large-scale rotary drum composting followed by Stack vermicomposting has the capacity to handle diverse organic waste sources in metropolitan locations. According to the literature, the approach promises a 70% decrease in total volume in 27 days with a nutrient-rich soil conditioner. A significant volume of organic waste can be bio-converted in diverse metropolitan locations employing a combined two-stage procedure. This decentralisation of trash management can improve municipal organisations' waste management efficiency while also reducing the pressure on dumpsites and landfills. Furthermore, decentralisation will result in employment development in the sector of waste management. When exposed to dry waste, resource recovery will become simpler, and thermal treatments will operate more efficiently.

## ACKNOWLEDGEMENT

We gratefully thank funding assistance from the Department of Science and Technology, Government of India (File No. DST/TDT/WMT/2017/153 (G)).

## REFERENCES

Alexander Cogut, 2016. Open Burning of Waste: A Global Health Disaster. R20 Regions of Climate Action. https://www.semanticscholar.org/paper/OPEN-BURNING-OF-WASTE%3A-A-GLOBAL-HEALTH-DISASTER-Cogut/9314094b71d64d20c79d9eba98d2d3d878f44c7f

Bernat, K., Kulikowska, D., Wojnowska-Baryła, I., Kamińska, A., 2022. Can the biological stage of a mechanical-biological treatment plant that is designed for mixed municipal solid waste be successfully utilized for effective composting of selectively collected biowaste? *Waste Manag.* 149, 291–301. doi:1016/J.WASMAN.2022.06.025

Biruntha, M., Karmegam, N., Archana, J., Karunai Selvi, B., John Paul, J.A., Balamuralikrishnan, B., Chang, S.W., Ravindran, B., 2020. Vermiconversion of biowastes with low-to-high C/N ratio into value added vermicompost. *Bioresour. Technol.* 297, 122398. doi:1016/j.biortech.2019.122398

Blossey, B., Schroeder, D., Hight, S.D., Malecki, R.A., 1994. Host specificity and environmental impact of the weevil *Hylobius transversovittatus*: A biological control agent of purple loosestrife (*Lythrum salicaria*). *Weed Sci.* 42, 128–133.

Chavan, D., Arya, S., Kumar, S., 2022a. Open dumping of organic waste: Associated fire, environmental pollution and health hazards. In: *Advanced Organic Waste Management*. Elsevier, Amsterdam, pp. 15–31. doi:1016/B978-0-323-85792-5.00014-9

Chavan, D., Lakshmikanthan, P., Mondal, P., Kumar, S., Kumar, R., 2019. Determination of ignition temperature of municipal solid waste for understanding surface and sub-surface landfill fire. *Waste Manag.* 97, 123–130. doi:1016/J.WASMAN.2019.08.002

Chavan, S., Yadav, B., Atmakuri, A., Tyagi, R.D., Wong, J.W.C., Drogui, P., 2022b. Bioconversion of organic wastes into value-added products: A review. *Bioresour. Technol.* 344, 126398. doi:1016/J.BIORTECH.2021.126398

Dash, S., Borah, S.S., Kalamdhad, A.S., 2021. Heavy metal pollution and potential ecological risk assessment for surficial sediments of *Deepor Beel*, India. *Ecol. Indic.* 122, 107265. doi:1016/J.ECOLIND.2020.107265

Deepthi, M.P., Kathireswari, P., Rini, J., Saminathan, K., Karmegam, N., 2021. Vermitransformation of monogastric *Elephas maximus* and ruminant *Bos taurus* excrements into vermicompost using *Eudrilus eugeniae*. *Bioresour. Technol.* 320, 124302. doi:1016/j.biortech.2020.124302

Devi, C., Khwairakpam, M., 2020. Bioconversion of Lantana camara by vermicomposting with two different earthworm species in monoculture. *Bioresour. Technol.* 296, 122308. doi:1016/J.BIORTECH.2019.122308

dos Muchangos, L.S., Tokai, A., 2020. Greenhouse gas emission analysis of upgrading from an open dump to a semi-aerobic landfill in *Mozambique*: The case of Hulene dumpsite. *Sci. African* 10, e00638. doi:1016/J.SCIAF.2020.E00638

El-Haggar, S.M., 2007. Current practice and future sustainability. *Sustain. Ind. Des. Waste Manag.* 1–19. doi:1016/B978-012373623-9/50003-4

FAO, 2019. Moving forward on food loss and waste reduction. *State Food Agric.* 2019. https://www.fao.org/3/ca6030en/ca6030en.pdf

Ganesh, K.S., Sridhar, A., Vishali, S., 2022. Utilization of fruit and vegetable waste to produce value-added products: Conventional utilization and emerging opportunities-A review. *Chemosphere* 287, 132221. doi:1016/J.CHEMOSPHERE.2021.132221

Gimžauskaitė, D., Aikas, M., Tamošiūnas, A., 2022. Recent progress in thermal plasma gasification of liquid and solid wastes. *Recent Adv. Renew. Energy Technol.* 155-196. doi:1016/B978-0-12-823532-4.00007-0

González, R., Hernández, J.E., Gómez, X., Smith, R., González Arias, J., Martínez, E.J., Blanco, D., 2020. Performance evaluation of a small-scale digester for achieving decentralised management of waste. *Waste Manag.* 118, 99–109. doi:1016/J.WASMAN.2020.08.020

Gupta, G., Datta, M., Ramana, G.V., Alappat, B.J., 2021. MSW incineration bottom ash (MIBA) as a substitute to conventional materials in geotechnical applications: A characterization study from India and comparison with literature. *Constr. Build. Mater.* 308, 124925. doi:1016/J.CONBUILDMAT.2021.124925

Hazarika, J., Khwairakpam, M., 2018. Evaluation of biodegradation feasibility through rotary drum composting recalcitrant primary paper mill sludge. *Waste Manag.* 76, 275–283. doi:1016/j.wasman.2018.03.044

Iyer, H., 2016. Case study of Mumbai: Decentralised solid waste management. *Procedia Environ. Sci.* 35, 101–109. doi:1016/J.PROENV.2016.07.054

Jain, M.S., Kalamdhad, A.S., 2018. Efficacy of batch mode rotary drum composter for management of aquatic weed (*Hydrilla verticillata* (L.f.) Royle). *J. Environ. Manage.* 221, 20–27. doi:1016/j.jenvman.2018.05.055

Jain, M.S., Kalamdhad, A.S., 2019. Drum composting of nitrogen-rich *Hydrilla Verticillata* with carbon-rich agents: Effects on composting physics and kinetics. *J. Environ. Manage.* 231, 770–779. doi:1016/j.jenvman.2018.10.111

Kalamdhad, A.S, Ali Khwairakpam, M., Kazmi, A.A., 2009a. Organic matter transformation during rotary drum composting. *Dyn Soil Dyn Plant* 3, 93–98.

Kalamdhad, A.S., Singh, Y.K., Ali, M., Khwairakpam, M., Kazmi, A.A., 2009b. Rotary drum composting of vegetable waste and tree leaves. *Bioresour. Technol.* 100, 6442–6450. doi:1016/j.biortech.2009.07.030

Kauser, H., Pal, S., Haq, I., Khwairakpam, M., 2020. Evaluation of rotary drum composting for the management of invasive weed *Mikania micrantha Kunth* and its toxicity assessment. *Bioresour. Technol.* 313, 123678. doi:1016/j.biortech.2020.123678

Kuznetsova, E., Cardin, M.A., Diao, M., Zhang, S., 2019. Integrated decision-support methodology for combined centralized-decentralized waste-to-energy management systems design. *Renew. Sustain. Energy Rev.* 103, 477–500. doi:1016/J.RSER.2018.12.020

Lévesque, J., Perreault, V., Mikhaylin, S., 2022. Food waste in a hotel foodservice: A case study identifying hot spots and strategies to prioritize towards a reduction. *Int. J. Gastron. Food Sci.* 100600. doi:1016/J.IJGFS.2022.100600

Lipton, W.V., 2004. Current issues in mixed waste management. *Chem. Health Saf.* 11, 28–30. doi:1016/j.chs.2003.10.001

Liu, X., Hu, C., Zhang, S., 2005. Effects of earthworm activity on fertility and heavy metal bioavailability in sewage sludge. *Environ. Int.* 31, 874–879. doi:1016/J.ENVINT.2005.05.033

Lo, A.Y., Liu, S., 2018. Towards sustainable consumption: A socio-economic analysis of household waste recycling outcomes in Hong Kong. *J. Environ. Manage.* 214, 416–425. doi:1016/J.JENVMAN.2018.03.029

Owusu-Ansah, P., Obiri-Yeboah, A.A., Nyantakyi, E.K., Woangbah, S.K., Yeboah, S.I.I.K., 2022. Ghanaian inclination towards household waste segregation for sustainable waste management. *Sci. African* 17, e01335. doi:1016/J.SCIAF.2022.E01335

Paul Choudhury, S., Dalasingh, B., Haq, I., Kalamdhad, A.S., 2021. Methane production and toxicity evaluation of petroleum refinery biosludge through optimization of different modes of heat. *Process Saf. Environ. Prot.* 154, 236–248. doi:1016/J.PSEP.2021.08.019

Pottipati, S., Chakma, R., Haq, I., Kalamdhad, A.S., 2022a. Composting and vermicomposting: Process optimization for the management of organic waste. In: Advanced Organic Waste Management. Elsevier, Amsterdam, pp. 33–43. doi:1016/B978-0-323-85792-5.00015-0

Pottipati, S., Kalamdhad, A.S., 2022. Thermophilic-mesophilic biodegradation: An optimized dual-stage biodegradation technique for expeditious stabilization of sewage sludge. *J. Environ. Manage.* 323, 116189. doi:1016/J.JENVMAN.2022.116189

Pottipati, S., Kundu, A., Kalamdhad, A.S., 2021. Process optimization by combining in-vessel composting and vermicomposting of vegetable waste. *Bioresour. Technol.* 126357. doi:1016/j.biortech.2021.126357

Pottipati, S., Kundu, A., Kalamdhad, A.S., 2022b. Performance evaluation of a novel two-stage biodegradation technique through management of toxic lignocellulosic terrestrial weeds. *Waste Manag.* 144, 191–202. doi:1016/J.WASMAN.2022.03.026

Rathore, P., Chakraborty, S., Gupta, M., Sarmah, S.P., 2022. Towards a sustainable organic waste supply chain: A comparison of centralized and decentralized systems. *J. Environ. Manage.* 315, 115141. doi:1016/J.JENVMAN.2022.115141

Rodić, L., Wilson, D., 2017. Resolving governance issues to achieve priority sustainable development goals related to solid waste management in developing countries. *Sustainability* 9, 404.

Saha, B., Devi, C., Khwairakpam, M., Kalamdhad, A.S., 2017. Vermicomposting and anaerobic digestion: Viable alternative options for terrestrial weed management: A review. *Biotechnol. Reports* 17, 70–76. doi:1016/j.btre.2017.11.005

Saha, J.K., Panwar, N., Singh, M.V., 2010. An assessment of municipal solid waste compost quality produced in different cities of India in the perspective of developing quality control indices. *Waste Manag.* 30, 192–201. doi:1016/J.WASMAN.2009.09.041

Schuster, S., Speck, M., van Herpen, E., Buchborn, F., Langen, N., Nikravech, M., Mullick, S., Eichstädt, T., Chikhalova, Y., Budiansky, E., Engelmann, T., Bickel, M., 2022. Do meal boxes reduce food waste from households? *J. Clean. Prod.* 375, 134001. doi:1016/J.JCLEPRO.2022.134001

Shah, A.V., Singh, A., Sabyasachi Mohanty, S., Kumar Srivastava, V., Varjani, S., 2022. Organic solid waste: Biorefinery approach as a sustainable strategy in circular bioeconomy. *Bioresour. Technol.* 349, 126835. doi:1016/J.BIORTECH.2022.126835

Sharma, D., Yadav, K.D., Kumar, S., 2018. Role of sawdust and cow dung on compost maturity during rotary drum composting of flower waste. *Bioresour. Technol.* 264, 285–289. doi:1016/j.biortech.2018.05.091

Sholokhova, A., Denafas, G., Mykhaylenko, V., 2022. Microplastics generation and concentration during mechanical-biological treatment of mixed municipal solid waste. *Environ. Res.* 214, 113815. doi:1016/J.ENVRES.2022.113815

Singh, W.R., Das, A., Kalamdhad, A., 2012. Composting of water hyacinth using a pilot scale rotary drum composter. *Environ. Eng. Res.* 17, 69–75.

Srigirisetty, S., Jayasri, T., Netaji, C., 2017. Open dumping of municipal solid waste: Impact on groundwater and soil. *Int. J. Curr. Eng. Sci. Res.* 4, 26–33.

Srivastava, V., Balakrishnan, M., 2022. Phytochemicals from fruit and vegetable waste generated in hotels: Optimization of recovery procedure and potential for value-addition. *Waste Manag.* 144, 401–410. doi:1016/J.WASMAN.2022.04.019

Trabold, T.A., Nair, V., 2018. Conventional food waste management methods. *Sustain. Food Waste-to-Energy Syst.* 29–45. doi:1016/B978-0-12-811157-4.00003-6

Varma, V.S., Kalamdhad, A.S., 2016. Efficiency of rotary drum composting for stabilizing vegetable waste during pre-composting and vermicomposting. *Environ. Process.* 3, 829–841. doi:1007/s40710-016-0181-z

Varma, V.S., Kalamdhad, A.S., Khwairkpam, M., 2016. Feasibility of Eudrilus eugeniae and Perionyx excavatus in vermicomposting of water hyacinth. *Ecol. Eng.* 94, 127–135. doi:1016/j.ecoleng.2016.05.058

Wei, Y., Wang, N., Lin, Y., Zhan, Y., Ding, X., Liu, Y., Zhang, A., Ding, G., Xu, T., Li, J., 2021. Recycling of nutrients from organic waste by advanced compost technology- A case study. *Bioresour. Technol.* 337, 125411. doi:1016/J.BIORTECH.2021.125411

Wilson, D.C., Rodic-Wiersma, L., Modak, P., Soós, R., Carpintero Rogero, A., Velis, C., Iyer, M., Simonett, O., 2016. *Global Waste Management Outlook, Global Waste Management Outlook.* United Nations.

Yadav, V., Bhurjee, A.K., Karmakar, S., Dikshit, A.K., 2017. A facility location model for municipal solid waste management system under uncertain environment. *Sci. Total Environ.* 603–604, 760–771. doi:1016/J.SCITOTENV.2017.02.207

# 9 Integration of Technologies for Value-Added Products from Industrial Waste in Ho Chi Minh City

*Le Thi Kim Oanh, Ho Thi Thanh Hien, Huynh Tan Loi, and Nguyen Dang Khoa*

## 9.1 INTRODUCTION

The industrial production value of Ho Chi Minh City (HCMC) accounts for about 45% of industrial output in the southern critical economic region and contributes about 16% of the national industrial production scale (DOIT, 2018). In HCMC, there are 17 Export Processing Zones (EPZs) and Industrial Zones (IZs) in operation with a total area of 3,306 ha, and the occupancy rate is 67% with a total of 1,160 enterprises in operation (HEPZA, 2020).

Industrial solid waste is generated from production activities, not belonging to the group of household wastes and hazardous wastes. According to Decree No. 08/2022/ND-CP promulgated on January 10, 2022, ash and slag are ordinary industrial solid waste, while foundry residue and sludge from wastewater treatment stations of EPZs/IZs are regulated as controlled waste. This means that these wastes must be strictly analyzed on a case-by-case basis so that they can be classified as industrial or hazardous waste. Thus, the promulgated decree has expanded opportunities to recover and recycle these specific solid wastes.

The sources of solid waste in general and ash–slag, sludge, and foundry residue, in particular, are mainly scattered and small, which are challenging to collect. Therefore, they have often been arbitrarily, unsystematically managed, and in a complex way. The recycling abilities of the industrial wastes mentioned above such as potential technologies, are still abandoned issues. Vietnam's Ministry of Natural Resources and Environment has concluded that the current industrial solid waste management system has not been applied according to the method of reduction, reuse, recycling, and energy recovery (MONRE, 2019).

Therefore, the determination of technical parameters related to industrial waste including the generation and norm, volume, composition; the current status of collection, transshipment, and transportation; the current status of reuse, treatment, and disposal; pointing out inadequacies in the management structure, systems of policy documents on solid waste management will contribute to completing the database set, thereby improving the efficiency of environmental management and resource recovery. Data on composition, characteristics, and emission factors for each type of waste is extremely useful and scientific basis for the design of industrial solid waste-related supply chains and is the foundation to contribute to the development of industrial symbiosis and a circular economy, meeting the United Nations' goal of 12 on sustainable development. In addition, this database will play an important role in the management and planning of industrial waste management infrastructure.

DOI: 10.1201/9781003386902-12

## 9.2 CURRENT STATUS OF INDUSTRIAL ACTIVITIES AND INDUSTRIAL WASTE MANAGEMENT IN HCMC, VIETNAM

The development orientation of IZs of HCMC up to 2025 is to transform existing IZs into green, clean gradually, and high-tech IZs. The city develops policies to attract investments in key industries, including mechanical engineering, electronics and informatics, chemicals and food processing. Other existing industries are focusing on textile dyeing, seafood processing, garment, and so on.

Industrial solid waste is collected and sorted at the source, and then the manufactory signs recycling and treatment contracts with operating companies. Some manufacturing factories are capable of on-site recycling of some types of waste, such as nylon and plastic, to produce raw plastic beads, returning to the production process. Parallelly, various waste recycling takes place in a complex manner, in particular, most concentrated in the suburbs of the city, such as Binh Tan District, District 12, District 9, District 2, District 7, District 8, Binh Chanh District, Cu Chi District, Nha Be District and Thu Duc District. These activities are concentrated on National Highway 1A, major roads and around industrial clusters, IZs and EPZs. In addition, there are many waste collection and recycling facilities operating without operating licenses. Types of collected waste for reuse and recycling are scraps of fabric, plastic, iron, aluminum, copper, paper, wood, lubricants, drums with chemicals/grease, and batteries. Non-hazardous solid waste that is reused and recycled accounts for about 70%–80% of the generated volume, the rest is disposed of by burning and burying together with domestic waste (DONRE, 2020).

Up to now, most EPZs/IZs have not built stations or points for industrial solid waste collection, storage, and treatment. The classification of industrial solid waste is done at each discharged source, but in the whole industrial zone and the whole city, there is no dedicated storage and treatment center/zone. Ash, slag, and foundry residue are rarely recycled and mainly end up in the landfill. However, the government's current orientation for this group of wastes is as a cement mineral additive, partly replacing clay in brick, concrete production, and ground leveling. Sludge has been gradually diverted from landfill solution to recycling solution to produce fertilizer by aerobic or anaerobic combined aerobic technology. Therefore, a large amount of industrial solid waste is stored in fields and vacant land, creating a large source of pollution, harmful to people, nature, and land use. Many electronic and chemical manufacturing enterprises need a centralized solid waste collection center, but due to many objective reasons, this requirement still needs to be met.

Currently, HCMC does not have a database set on composition, characteristics, generation rate, and technological solutions for recycling industrial waste. Specifically, the source of the database on ash, slag, sludge, and mold is an urgent need to support the state management of industrial waste in HCMC.

In general, some characteristics to keep in mind for an industrial solid waste management system in IZs are the following:

- Most waste sources are small in volume and dispersed, resulting in high transportation and disposal costs. In many cases, different types of waste are collected and stored together, making it difficult to recycle and treat. There has not been a good investment in waste storage and classification areas at the discharged sources.
- The composition of industrial solid waste is diverse and fluctuates wildly, depending on input materials, technologies, operating techniques, and so on. It affects the qualitative determination of hazardous or non-hazardous recycling processes and industrial symbiosis opportunities.
- Industrial waste has yet to be monitored and managed thoroughly from the discharged sources to the final disposal. In detail, within the IZs, waste is managed and monitored by HEPZA (Ho Chi Minh City Export Processing and Industrial Zones Authority). When transporting waste out of the industrial park to the recycling and treatment plant, the waste is monitored by HEPA (Ho Chi Minh City Environmental Protection Agency), and

environmental policies at the local and national levels. There has not been an industrial solid waste management system that can monitor the flow of waste from the sources to the final treatment site.

- The choice of recycling or disposal technology is optional, depending on the market mechanism. Therefore, the information/data related to the treatment technology and the amount of waste treated by each technology is not published.
- Information on generation sources, the volume of each type of waste, and waste composition are only recorded through waste collection and treatment contracts. The management unit has yet to announce the composition and codes of waste, as well as the regulations guiding the selection of optimal waste recycling and treatment technology.
- The issued management policy has actively supported waste management, recycling, and industrial symbiosis activities. However, when it comes to implementation, there need to be more documents guiding stakeholders in the management of industrial solid waste at the source in the collection, transportation, recycling, and industrial symbiosis process.
- The Ministry of Natural Resources and Environment has assessed that "solid waste management and treatment in our country has neither been applied in the past time according to the integrated management method, nor paying attention to 5 Rs solutions such as reduce, reuse, recycle and energy recovery".

## 9.2.1 Current Circular Economy Development in Industrial Zones in HCMC

Figure 9.1 described by Geissdoerfer et al. (2020) was about the circular economy, including prolonging product life, reuse, remanufacturing/refurbishing, and recycling. In industrial zones, where the products are produced, the aspects of reuse, remanufacturing, and recycling are considered.

### 9.2.1.1 Technical Aspects

Reusing, remanufacturing, and recycling industrial waste are being adopted profoundly in developing countries like Vietnam. These activities are carried out on a small scale by the private sector. In HCMC, investment in large-scale storage, sorting, recycling, and treatment are limited.

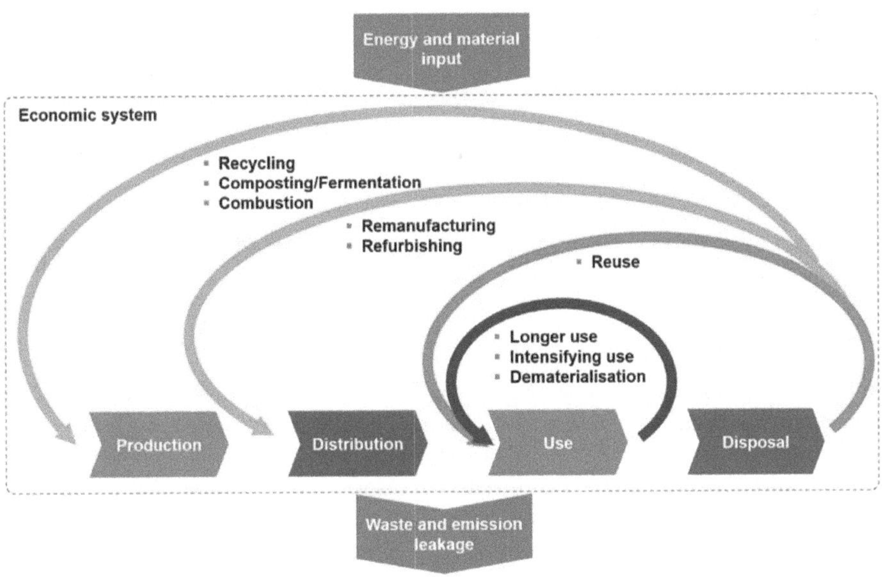


**FIGURE 9.1** Circular economy. (*Source*: Geissdoerfer et al., 2020.)

These small-scale recycling activities have not been controlled in terms of product quality. Therefore, the quality of products after recycling is often reduced by one or more orders compared to the quality of the input waste materials. Therefore, for the purpose of more extended use of this low-quality recycled product, it will also be limited. Due to low investment capital, small-scale units apply outdated technologies and equipment, which reduces product quality and pollutes the environment. Besides, the treatment of pollution arising from recycling activities has neither been controlled by investors nor monitored well by the authorities.

#### 9.2.1.2 Management Aspects

The role of stakeholders in the entire state management system has not been pointed out to improve the efficiency of reuse, remanufacturing, recycling, and prolonging the product life in most industries. Industrial plants with activities to classify and recover resources from industrial waste are rare cases, for which are unreported and unaccounted. For example, as a reputable brand, IKEA has built specific strategies, plans, and indicators to control waste management activities toward sustainable development (IKEA, 2021). To meet the requirements of IKEA, all processing plants, that were having a contract with IKEA, must comply with the planning and periodic monitoring and evaluation of waste management and reduction activities. Thus, IKEA's management policy has contributed to promoting resource saving, minimizing waste at source, classifying and storing waste properly, and contributing to increasing the efficiency of reuse and recycling activities. However, IKEA does not control the reuse and recycling of waste when the collection vehicle leaves IKEA waste sources. Therefore, whether the waste is recycled by the technological requirements is still a question.

### 9.2.2 Waste Management Policy and Promotion of Circular Economy Development

In aspects affecting the effective implementation of the circular economy model in Vietnam, the policy system is relatively complete to promote industrial waste recycling and industrial symbiosis. Besides waste management and pollution reduction policies, there are also many groups of policies, regulations, and tools to improve recycling efficiency.

In HCMC, in addition to the implementation of the Environmental Law and the government's general policies on waste management, the city has issued many encouraging and supportive documents to enhance the efficiency of the recycling industry and industrial symbiosis. However, the document system still needs to be completed, overlapping, and not fully implemented. The implementation guidance still has many problems that need to be more timely and specific. Specific mechanisms for incentives for waste recycling, reuse, and energy recovery from solid waste treatment need to be improved and consistent. Therefore, it is difficult for small businesses to access these incentives (while most recyclers are small scale). This also causes difficulties in the operation of the business.

## 9.3 POTENTIAL OF SOME SPECIFIC INDUSTRIAL WASTES TO REUSE, RECYCLING, AND INDUSTRIAL SYMBIOSIS

### 9.3.1 Ash and Slag

Due to the low labor cost, recycling activities are not strictly controlled both in terms of product quality and the management of pollution emissions into the environment (Salhofer et al., 2021). Therefore, recycling activities in developing countries, both in the residential and industrial areas, are developing strongly. The proportion of industrial wastes with the potential for reuse, remanufacturing and recycling is high (Schneider et al., 2017). For industrial waste in HCMC, common types of industrial waste that are reused, remanufactured, and recycled include rags, waste plastic, plastic/metal containers, waste rubber, wood chips/waste wood, agricultural by-products, non-hazardous industrial sludge, biodegradable organic waste, solvents, wastewater with high BOD (biogas recovery)...Some other types of industrial waste such as ash and slag, mold waste, industrial sludge from centralized wastewater treatment stations, and so on, have not been reused, remanufactured, or recycled widely due to their high specificity and high contamination variation, easily mistaken as hazardous waste. From January 2022, Circular 02/BTNMT of the Ministry of Natural Resources and Environment of Vietnam has regulated ash and slag as common waste, promoting reuse and recycling activities.

According to statistics of HEPZA (2018), there are currently more than 200 fuel-burning furnaces operating in the IZs of HCMC. The capacity of the furnaces used is mainly small and medium; even due to the specificity, some production facilities use furnaces with minimal capacities. These sources of ash and slag emissions are scattered and small, making it difficult to collect and store. In many waste sources, these wastes are collected and stored with other waste. They are even collected with hazardous waste; therefore, they are not recycled but treated as hazardous waste.

Among the fuels used, coal, diesel oil, cashew shells, firewood, and sawdust are the most commonly used by 131 production facilities at 17 IZs of HCMC, with 22.1%, 18.3%, 16.8%, 14.5%, and 9.9%, respectively. In particular, the amount and rate of fuel used commonly at 17 IZs of HCMC in 2020 are DO: 110,507 tons/month (84.99%), coal: 11,757 tons/month (9.04%); sawdust: 2,025 tons/month (1.56%), firewood: 2,016 tons/month (1.55%) and cashew shells: 1,200 tons/month (0.92%).

#### 9.3.1.1 pH

Depending on the pH value and calcium/sulfur ratio, ash is classified into acid (pH 1.2–7), slightly alkaline (pH 8–9), and strongly alkaline ash (pH 11–13) (Bhatt et al., 2019). Table 9.1 shows that the pH of all samples at IZs of HCMC are either slightly alkaline or strongly alkaline, with some samples having pH values exceeding Vietnam's standard for hazardous waste thresholds.

#### 9.3.1.2 Heavy Metals

The analysis results of heavy metal indicators showed that the samples of fly ash and slag met QCVN 07:2009/BTNMT on the hazardous waste threshold. Tables 9.2 and 9.3 show the heavy metal and radioactive composition of fly ash and bottom ash/slag from various sources.

#### 9.3.1.3 Inorganic Compositions and Characteristics

All samples of ash and slag from burning fuels including coal, pellets, firewood, and cashew shells met the restriction factor requirement specified in Appendix IV of QCVN 01-189:2019/BNNPTNN (Vietnam's standard on fertilizer quality). Moreover, the samples meet most of the requirements for reuse for ground leveling materials according to TCVN 12249:2018 with most parameters, except for the parameter of "Safe radioactive αβ". However, the requirements for reuse for the production of concrete, mortar, and cement according to TCVN 10302:2014, as well as for reuse as a mineral additive for roller compacted concrete according to TCVN 8825:2011 could not be satisfied. Tables 9.4 and 9.5 summarize the compositions and characteristics considered in Vietnam's standard for various construction materials, and conditions for fertilizer and ground leveling purposes.

**TABLE 9.1**
**pH in Ash from Different Fuel Sources**

| Fuel | pH | Number of Sample | QCVN 07:2009/ BTNMT |
|---|---|---|---|
| Coal | 10.3–11.4 | $n = 2$ | pH ≤ 2.0 |
| Firewood | 8.1–12.5 | $n = 3$ | pH ≥ 12.5 |
| Cashew shells | 11.7–13.1 | $n = 3$ | |
| Sawdust | 12.2–12.7 | $n = 2$ | |

*QCVN 07:*2009/BTNMT: Vietnam's standard for hazardous waste thresholds.

**TABLE 9.2**
**The Heavy Metal and Radioactive Composition of Fly Ash From Various Sources**

| Parameters | Fly Ash from Coal | Fly ash from Fuel Pellets | Fly Ash from Firewood | Fly Ash from Others Biomass | TCVN 12249:2018 | | QCVN 01-189:2019/ BNNPTNN | QCVN 07:2009/ BTNMT |
|---|---|---|---|---|---|---|---|---|
| | *n* = 5 | *n* = 6 | *n* = 7 | *n* = 3 | Colum A: | Colum B: | | |
| Safe radioactive αβ | 0.66–1.84 | 1.17–1.99 | 0.28–1.69 | 0.78–2.15 | ≤1 | | | |
| | 0.09–0.27 | 0.17–0.30 | 0.04–0.25 | 0.11–0.32 | | | | |
| As (mg/L) (LOD = 0,015) | Nd | Nd–0.290 | Nd | Nd | 0.05 | 0.1 | ≤10 | 2 |
| Cd (mg/L) (LOD = 0,005) | Nd | Nd | Nd | Nd | 0.05 | 0.1 | ≤5 | 0,5 |
| Cu (mg/L) (LOD = 0,005) | Nd–0.015 | Nd–0.025 | Nd | Nd | 2 | 2 | | - |
| Hg (mg/L) (LOD = 0,001) | Nd | Nd | Nd | Nd | 0.005 | 0.01 | ≤2 | 0,2 |
| Ni (mg/L) (LOD = 0,005) | 0.006–0.097 | Nd–0.013 | Nd | Nd | 0.2 | 0.5 | | 70 |
| Pb (mg/L) (LOD = 0,005) | Nd | Nd–0.016 | Nd | Nd | 0.1 | 0.5 | ≤200 | 15 |
| Zn (mg/L) (LOD = 0,005) | Nd–0.623 | Nd–0.014 | Nd | Nd | 3 | 3 | | 250 |

*TCVN 12249:*2018: Vietnam's standard on the use of ash and slag from coal-fired thermal power plants as ground leveling materials.
*Column A indicates the values of parameters of leaching water when it is discharged into the water sources serving tap water supply.*
*Column B indicates the values of parameters of leaching water when it is discharged into the water sources not serving tap water supply.*
*QCVN 01-189:*2019/BNNPTNN: Vietnam's standard on fertilizer quality.
*QCVN 07:*2009/BTNMT: Vietnam's standard for hazardous waste thresholds.
*Nd, Not detected.*

**TABLE 9.3**
**The Heavy Metal and Radioactive Composition of Bottom Ash/Slag from Various Sources**

| Parameters | Bottom Ash from Coal | Bottom Ash from Fuel Pellets | Bottom Ash from Firewood | Bottom Ash from Others Biomass | TCVN 12249:2018 | | QCVN 01-189:2019/ BNNPTNN | QCVN 07:2009/ BTNMT |
|---|---|---|---|---|---|---|---|---|
| | ($n$ =5) | ($n$ =5) | ($n$ =7) | ($n$ =3) | Column A | Column B | | |
| Safe radioactive | 0.22–1.67 | 0.24–0.29 | 1.08–2.25 | 0.39–0.67 | ≤1 | | | |
| | 0.03–0.24 | 0.03–0.04 | 0.16–0.33 | 0.06–0.10 | | | | |
| As (mg/L) (LOD = 0.015) | Nd | Nd | Nd | Nd | 0.05 | 0.1 | ≤10 | 2 |
| Cd (mg/L) (LOD = 0.005) | Nd | Nd | Nd | Nd | 0.05 | 0.1 | ≤5 | 0.5 |
| Cu (mg/L) (LOD = 0.005) | Nd | Nd–0.014 | Nd | Nd–0.007 | 2 | 2 | | - |
| Hg (mg/L) (LOD = 0.001) | Nd | Nd | Nd–0.018 | Nd–0.063 | 0.005 | 0.01 | ≤2 | 0.2 |
| Ni (mg/L) (LOD = 0.005) | 0.032–0.039 | Nd–0.019 | Nd–0.127 | Nd–0.036 | 0.2 | 0.5 | | 70 |
| Pb (mg/L) (LOD = 0.005) | Nd | Nd | Nd | Nd | 0.1 | 0.5 | ≤200 | 15 |
| Zn (mg/L) (LOD = 0.005) | Nd–0.397 | Nd–0.248 | Nd–0.347 | Nd–0.023 | 3 | 3 | | 250 |

*TCVN 12249:*2018: Vietnam's standard on the use of ash and slag from coal-fired thermal power plants as ground leveling materials.
*Column A indicates the values of parameters of leaching water when it is discharged into the water sources serving tap water supply.*
*Column B in indicates the values of parameters of leaching water when it is discharged into the water sources not serving tap water supply.*
*QCVN 01-189:*2019/BNNPTNN: Vietnam's standard on fertilizer quality.
*QCVN 07:*2009/BTNMT: Vietnam's standard for hazardous waste thresholds.
*Nd, Not detected.*

**TABLE 9.4**

**Inorganic Compositions and Characteristics of Fly Ash from Various Sources**

| ParaMeters | Fly Ash from Coal | Fly Ash from Fuel Pellets | Fly Ash from Firewood | Fly Ash from Others Biomass | TCVN 10302:2014 | | | | TCVN 10302:2014 | | TCVN 8825:2011 | | TCVN 6882:2016 | | |
|---|---|---|---|---|---|---|---|---|---|---|---|---|---|---|---|
| | ($n$ = 5) | ($n$ = 6) | ($n$ = 7) | (n = 3) | $C\text{-}_a$ | $C\text{-}_b$ | $C\text{-}_c$ | $C\text{-}_d$ | F | C | Natural Activated Admixture | Artificial Activated Admixture | Filling Admixture | Activated Admixture | Filling Admixture |
| $K_2O$ (%) | 1.1–16.7 | 4.7–10.6 | 0.8–18.5 | 5.8–49.7 | | | | | | | | | | | |
| $CaO_{free}$ (%) | 12.8–54.3 | 34.0–64.8 | 1.6–63.7 | 9.7–47.5 | ≤2 | ≤4 | ≤4 | ≤2 | ≤1 | ≤3 | | | | | |
| $SO_3$ (%) | 2.9–3.7 | 1.0–10.8 | 0.2–2.7 | 2.0–12.9 | ≤5 | ≤5 | ≤6 | ≤3 | ≤3.5 | ≤5 | ≤4.0 | ≤5.0 | ≤4.0 | | |
| $SiO_2$ (%) | 3.8–42.0 | 8.2–17.1 | 2.5–91.3 | 3.5–12.6 | | | | | | | | | | | |
| $Al_2O_3$ (%) | 1.4–19.9 | 1.8–3.7 | 0.2–11.5 | 1.3–3.6 | | | | | | | | | | | |
| $Fe_2O_3$ (%) | 2.3–15.1 | 3.9–6.6 | 0.9–37.8 | 3.5–5.1 | | | | | | | | | | | |
| Total $SiO_2 + Al_2O_3 + Fe_2O_3$ (%) | 7.5–77.0 | 14.9–27.4 | 4.0–92.4 | 8.3–19.8 | ≥45 | | | | | | ≥70 | ≥70 | - | | |
| $Cl^-$ (%) | 0.03–2.17 | 0.46–8.54 | 0.02–3.64 | 2.04–4.23 | ≤0.1 | - | - | ≤0.1 | | | | | | | |
| Loss on ignition (%, max) | 2.7–22.8 | 2.6–74.7 | 2.3–46.2 | 1.7–3.8 | ≤5 | ≤9 | ≤7 | ≤5 | ≤8* | ≤6 | ≤10 | ≤6 | - | | |
| Moisture content (%, max) | 0.3–18.5 | 0.5–61.3 | 0.7–32.7 | 0.4–2.1 | ≤3 | | | | ≤1 | ≤1 | ≤3 | ≤3 | ≤3 | | |
| Cumulation retained on sieve size 45 μm (%) | 24.5–52.6 | 68.3–94.5 | 52.0–87.3 | 72.9–85.0 | ≤25 | ≤34 | ≤40 | ≤18 | | | ≤34 | ≤34 | - | | |
| Required amount of water compared to control sample (%) | 109.5–110.5 | 104.3–111.8 | 108.2–110.1 | 99.2–111.4 | ≤105 | | | | | | ≤115 | ≤105 | ≤115 | | |
| Expansion in autoclave (%) | 0.08–0.09 | 0.02–0.17 | 0.06–0.23 | 0.03–0.08 | | | | | | | ≤0.8 | | | | |
| Total Na eq. Alk, ($\%Na_2O_{td}$) | 0.33–1.16 | 0.37–7.26 | 0.11–1.23 | 1.83–16.33 | ≤1.5 | | | | | | | | | | |

*(Continued)*

**TABLE 9.4 (*Continued*)**

**Inorganic Compositions and Characteristics of Fly Ash from Various Sources**

| ParaMeters | Fly Ash from Coal | Fly Ash from Fuel Pellets | Fly Ash from Firewood | Fly Ash from Others Biomass | TCVN 10302:2014 | | | | TCVN 10302:2014 | | TCVN 8825:2011 | | TCVN 6882:2016 | | |
|---|---|---|---|---|---|---|---|---|---|---|---|---|---|---|---|
| | (*n* = 5) | (*n* = 6) | (*n* = 7) | (n = 3) | C-[a] | C-[b] | C-[c] | C-[d] | F | C | Natural Activated Admixture | Artificial Activated Admixture | Filling Admixture | Activated Admixture | Filling Admixture |
| Compressive strength at 7 days (%) | 49.9–72.5 | 54.4–82.9 | 45.8–88.7 | 44.7–84.2 | | | | | | | ≥75 | | - | ≥75 | - |
| Compressive strength at 28 days (%) | 51.1–83.5 | 63.1–88.3 | 49.7–97.5 | 56.2–100.8 | | | | | | | ≥75 | ≥75 | - | ≥75 | - |
| Natural radioactivity Aeff (Bq/kg) | 110–4587 | 2329–5233 | 167–3284 | 1044–5687 | ≤370 (civil construction)<br>≤740 (industrial and road construction) | | | | | ≤370 | | | | | |
| Time to finish setting of lime mortar–mineral additives (h) | 14–20 | 12.0–42.0 | 18–46 | 14–18 | | | | | | | | | | ≤96 | - |
| Soluble alkali content of mineral additives after 28 days (%) | 0.33–1.16 | 0.37–7.26 | 0.11–1.23 | 1.83–16.33 | | | | | | | | | | | ≤1.5 |

*TCVN 10302:*2014: Vietnam's standard on Fly ash active additives for concrete & mortar.

*Basal ash ©:* CaO content >10%.

*When burning anthracite coal, fly ash can be used with the corresponding burning loss:* field c to 12%; field d to 10%, by agreement or by accepted test results.

*TCVN 10302:*2014: Vietnam's standard on Fly ash active additives for cement.

*Acid ash (F):* CaO content up to 10%.

*Basal ash ©:* CaO content >10%.

*When burning anthracite coal, fly ash with a loss of up to 12% on heating may be used by agreement or by accepted test results.*

*TCVN 8825:*2011: Vietnam's standard on Mineral additive for roller compacted concrete.

*TCVN 6882:*2016: Vietnam's standard on Mineral additives for cement.

[a] *For heavy and light armored concrete.*

[b] *For heavy and light non-armored concrete.*

[c] *For porous concrete.*

[d] *For concrete in special conditions.*

**TABLE 9.5**
**Inorganic Compositions and Characteristics of Bottom Ash from Various Sources**

| | | | | | TCVN 8825:2011 | | | TCVN 6882:2016 | |
|---|---|---|---|---|---|---|---|---|---|
| Parameters | Bottom Ash from Coal ($n = 5$) | Bottom Ash from Fuel Pellets ($n = 5$) | Bottom ash from Firewood ($n = 7$) | Bottom Ash from Other Biomass ($n = 3$) | Natural Activated Admixture | Artificial Activated Admixture | Filling Admixture | Activated Admixture | Filling Admixture |
| $K_2O$ (%) | 1.3–37.1 | 2.2–11.9 | 6.8–48.0 | 5.9–35.9 | | | | | |
| $CaO_{free}$ (%) | 6.7–15.3 | 0.9–46.5 | 11.4–54.0 | 10.9–38.2 | | | | | |
| $SO_3$ (%) | 1.3–4.6 | 0.1–1.6 | 1.2–1.8 | 0.6–1.8 | ≤4.0 | ≤5.0 | ≤4.0 | | |
| $SiO_2$ (%) | 19.5–39.0 | 17.4–94.2 | 11.4–45.2 | 11.7–53.4 | | | | | |
| $Al_2O_3$ (%) | 1.7–21.6 | 0.5–2.9 | 1.4–6.8 | 2.4–8.6 | | | | | |
| $Fe_2O_3$ (%) | 4.1–41.9 | 0.4–4.9 | 2.2–5.2 | 2.8–11.7 | | | | | |
| Total $SiO_2 + Al_2O_3 + Fe_2O_3$ (%) | 25.3–80.2 | 25.2–95.1 | 15.4–54.2 | 18.0–73.7 | ≥70 | ≥70 | - | | |
| $Cl^-$ (%) | 0.02–0.10 | 0.38–0.38 | 0.06–0.41 | 0.02–0.98 | | | | | |
| Loss on ignition (%, max) | 5.5–25.0 | 24.2–51.3 | 17.6–68.7 | 8.3–55.9 | ≤10 | ≤6 | - | | |
| Moisture content (%, max) | 3.1–18.9 | 1.6–39.4 | 13.6–51.0 | 6.8–43.2 | ≤3 | ≤3 | ≤3 | | |
| Cumulation retained on sieve size 45 μm (%) | 93.6–99.7 | 96.1–97.3 | 83.1–99.2 | 89.1–97.4 | ≤34 | ≤34 | - | | |
| Required amount of water compared to control sample (%) | 98.4–105.8 | 101.5–104.8 | 99.4–109.9 | 99.6–100.4 | ≤115 | ≤105 | ≤115 | | |
| Expansion in autoclave (%) | 0.05–0.16 | 0.10–0.11 | 0.06–0.25 | 0.07–0.13 | ≤0.8 | ≤0.8 | ≤0.8 | | |
| Total Na eq. Alk, (%$Na_2O_{td}$) | 0.02–8.05 | 0.06–2.31 | 0.29–9.91 | 0.06–0.47 | ≤1.5 | ≤1.5 | ≤1.5 | | |
| Compressive strength at 7 days (%) | 61.8–98.1 | 84.4–93.9 | 49.0–52.9 | 65.9–89.6 | ≥ 75 | ≥75 | - | ≥75 | - |
| Compressive strength at 28 days (%) | 66.5–106.4 | 104.8–105.0 | 52.0–67.6 | 85.5–98.3 | ≥75 | ≥75 | - | ≥75 | - |

*(Continued)*

**TABLE 9.5 (*Continued*)**
**Inorganic Compositions and Characteristics of Bottom Ash from Various Sources**

| Parameters | Bottom Ash from Coal ($n$ = 5) | Bottom Ash from Fuel Pellets ($n$ = 5) | Bottom ash from Firewood ($n$ = 7) | Bottom Ash from Other Biomass ($n$ = 3) | TCVN 8825:2011 | | | TCVN 6882:2016 | |
|---|---|---|---|---|---|---|---|---|---|
| | | | | | Natural Activated Admixture | Artificial Activated Admixture | Filling Admixture | Activated Admixture | Filling Admixture |
| Natural radioactivity Aeff (Bq/kg) | 58–447 | 62–74 | 279–580 | 105–183 | | | | | |
| Time to finish setting of lime mortar–mineral additives (h) | 22–48 | 18–40 | 18–40 | 18–40 | | | | ≤96 | - |
| Soluble alkali content of mineral additives after 28 days (%) | 0.02–8.05 | 0.06–2.31 | 0.29–9.91 | 0.06–0.47 | | | | | ≤1.5 |

*TCVN 8825*:2011: Vietnam's standard on mineral additive for roller compacted concrete.
*TCVN 6882*:2016: Vietnam's standard on mineral additives for cement.

### 9.3.2 Sludge from Wastewater Treatment Plants

Based on the characteristics of the above five production types, sludge is classified as non-hazardous and potentially hazardous. Potentially hazardous sludge groups include paper, recycled paper, paper packaging, and packaging printing. The group of non-hazardous sludge includes processing agricultural products, food, beverage, seafood processing, and animal feed processing. The sludge composition is illustrated in Table 9.6.

The comparison results show that the sludge compositions of the agricultural products, food and beverage processing; seafood processing; and animal feed processing do not exceed the hazardous thresholds specified in QCVN 07:2009/BTNMT and QCVN 50:2013/BTNMT. This is the basis for proposing solutions to recycle and reuse this group of sludge. Compared with the Vietnam national standard on fertilizer quality (QCVN 01-189:2019/BNNPTNT), the sludge of the non-hazardous group is eligible to be used as fertilizer. The values of P, K, and minerals are low and can be easily overcome in fertilizer production. Besides, sewage sludge can be mixed with other wastes such as ash, slag, and molds to balance the composition in the mixture, enhancing the recyclability and efficiency of recycling.

For the "potentially hazardous" sludge group, compared with QCVN 50: 2013 on the hazardous threshold for sludge from water treatment, the analysis results of the sludge sample showed that only one criterion exceeded the threshold, which is chromium. The results show that only 1 out of 6 factories belong to the paper processing/paper recycling/packaging industries, and only 1 out of nine fact factories belonging to the group of packaging printing have sludge samples whose Cr concentration exceeds the hazardous threshold. Based on these results, it is strongly recommended to analyze the Cr composition in the sludge before recycling and reusing the sludge belonging to these two groups of industries. It is necessary to have more in-depth studies on the technology and chemicals used by the wastewater treatment plants to determine the correlation with the properties of the sludge sample, from which to develop appropriate management regulations according to the characteristics.

## 9.4 TECHNICAL FACTORS TO PROMOTE CIRCULAR ECONOMY IN INDUSTRIAL ZONE

So far, in the world and Vietnam, there have been many applications in the reuse, recycling, and industrial symbiosis of ash, slag, and sludge waste. Typical applications are leveling materials, road filling materials, construction materials, fertilizers, agricultural soil amendments, and forest land. In addition, many applications are in the pilot stage or are the results of scientific research.

### 9.4.1 Reuse and Recycle Industrial Wastes

According to the available and reference technologies in reusing and recycling industrial wastes, this chapter has aggregated data as shown in Tables 9.7 and 9.8.

### 9.4.2 Industrial Symbiosis

In addition to the reuse and recycling of each type of waste, as mentioned in the above sections, Vietnam is promoting industrial symbiotic solutions to allow materials to be used more sustainably and contribute to the creation of a circular economy. Industrial symbiosis is an operational process in which wastes or by-products of industrial activities are used or used in combination as raw materials for the production process (Domenech et al., 2019). The symbiotic waste mixture has the ability to balance the waste components, helping the composition of the waste mixture not to exceed the hazardous threshold and meet the requirements for recycled materials.

**TABLE 9.6**
**Sludge Composition from Wastewater Treatment Systems of $in IZs of HCMC**

| Parameters | Units | Processing Agricultural Products, Food, Beverage | Seafood Processing | Animal Feed Processing | Paper, Recycled Paper, Paper Packaging | Packaging Printing | QCVN 07:2009/ BTNMT | QCVN 50:2013/ BTNMT |
|---|---|---|---|---|---|---|---|---|
| pH | - | 7.00–8.41 | 5.37–8.53 | 6.71–8.53 | 7.21–8.1 | 5.35–7.51 | pH ≥12.5<br>pH ≤2.0 | - |
| Dry matter | % | 0.39–24.75 | 4.33–46.39 | 1.83–20.45 | 9.72–88.62 | 6.61–34.44 | - | - |
| Moisture content | % | 75.3–99.6 | 53.6–95.7 | 79.6–98.2 | 11.4–90.3 | 65.6–93.4 | - | - |
| Organic matter | % | 34.4–95.3 | 46.0–97.7 | 45.9–68.9 | 2.2–52.7 | 35.5–96.2 | - | - |
| T-N | % | 0.9–5.54 | 3.06–5.44 | 3.17–5.44 | 0.15–1.2 | 0.13–2.23 | - | - |
| T-P | % | 0.1–4.31 | 0.22–4.38 | 1.36–4.38 | 0.03–0.18 | 0.02–0.36 | - | - |
| As<br>(LOD = 0.3 mg/kg) | mg/kg | 0.41–15.26 | 0.86–3.58 | Nd | Nd | 2.02–3.63 | 40 | 40 |
| Cd | mg/kg | - | - | - | 0.2–0.2 | 0.81–1.18 | 10 | 10 |
| Cr | mg/kg | - | - | - | 6.5–140.7 | 19.6–115.4 | 100 | 100 |
| Hg<br>(LOD = 0.1 mg/kg) | mg/kg | - | - | - | Nd | Nd | 4 | 4 |

*Note:*
*Nd:* not detected
-: No data

**TABLE 9.7**
**Summary of Current Solutions to Reuse and Recycle Ash and Slag**

| No. | Solutions | Explanations |
|---|---|---|
| 1 | Application of fly ash in ground leveling and road leveling activities | Many countries worldwide have commonly used fly ash and even slag to level the ground, raise elevations, and build roads. This activity helps save soil leveling resources, and limit pollution from soil transportation activities.<br>In Vietnam, the demand for leveling is very high. However, the application of ash in leveling is still limited due to legal problems related to the hazardous level of ash. Circular 02/2022 of Vietnam has classified ash and slag from industrial production as ordinary industrial waste (not hazardous waste). This will create an incentive to reuse and recycle this type of waste. |
| 2 | Application of fly ash to make fertilizer, soil improver | In the composition of fly ash contains many minerals, which can be added to the composition of fertilizers. Besides, fly ash can also create the soil's structure to suit the user's needs. Reduce 1 ton of $CO_2$ when using 38.18 tons of fly ash as a soil conditioner (Montes-Hernandez et al., 2009).<br>In Vietnam, the use of fly ash as a raw material for fertilizer production is still minimal due to obstacles related to regulations on determining whether fly ash is a hazardous waste or not. In addition, there need to be more guidelines for using fly ash as a raw material in fertilizer production.<br>Research results of Oanh et al. (2022) have demonstrated that the quality of fly ash and slag from all furnaces with different fuels is satisfactory for fertilizer production. |
| 3 | Producing bricks from fly ash | Recycling fly ash to produce bricks has been very popular around the world. In Vietnam, it has also been applied to the source of fly ash from large-scale coal burning. Fly ash from biomass burning has not yet been applied.<br>Producing bricks from fly ash helps to reduce $CO_2$ emissions (50%–60%), save water (10%–20%), and save clay resources. Brick products from fly ash are light and have poorer thermal conductivity (Nihalani et al., 2020). |
| 4 | Producing concrete from fly ash | Fly ash is widely used to produce concrete, especially lightweight concrete, aerated autoclaved concrete (AAC). In Vietnam, fly ash from biomass burning does not meet the standards for the production of concrete.<br>It reduces 2 tons of $CO_2$ for every ton of fly ash to replace cement. Fly ash replaces 1/3–1/4 of the total amount of raw materials for cement production (Nihalani et al., 2020). Fly ash concrete products are lightweight, hard, and durable. This is a superior sound and heat insulation product.<br>Research by Oanh et al. (2022) has demonstrated that the quality of fly ash from different sources (coal, fuel pellets, firewood, biomass) has a big difference. There are some parameters in the composition of the ash that may not meet the requirements of brick or concrete production according to Vietnam'construction standards. |

**TABLE 9.8**
**Summary of Current Solutions to Reuse and Recycle Sludge**

| No. | Solutions | Explanations |
|---|---|---|
| 1 | Application of sludge to make compost/ fertilizer | Vietnam has recycled sludge to make compost/ fertilizer. Sludge from wastewater treatment plants of industrial plants and IZs has a high organic content, high N, P, and minerals nutrient content, which is very suitable for fertilizer production requirements. However, in many cases, sludge contains hazardous components (heavy metals, toxic chemicals, and so on), so it has affected the efficiency of the reuse and recycling of our resources.<br>Research by Oanh et al. (2022) shows that the sludge from the wastewater treatment plant of the packaging printing industry has a chromium concentration exceeding the hazardous standard. Sludge from wastewater treatment plants of other industries is satisfactory for fertilizer production. |
| 2 | Application of sludge to produce biogas | Waste sludge treatment by anaerobic digestion technology to produce biogas has been widely applied in many countries around the world. Biogas is converted mainly into heat and partly into electricity. In Vietnam, low heat demand and low electricity prices do not have a solid incentive to apply this technology. In Vietnam, there have been many studies on anaerobic sludge treatment; however, so far, no plant has officially operated. There is currently one pilot project in HCMC.<br>In the context of the global energy crisis and the implementation of the government's sustainable development goal, anaerobic technology will be an effective solution. |
| 3 | Application of sludge in ground leveling and road leveling activities | Waste sludge contains high organic matter and moisture content, making it challenging to replace sand in ground-level operations. Besides, the sludge may contain pollutant components affecting the leveling area. Therefore, sludge from the wastewater treatment plant is treated before application for leveling purposes.<br>In Vietnam, land-leveling activities have taken place on a very small scale and are not controlled in terms of environmental safety. |

#### 9.4.2.1 Industrial Symbiosis Solution for Fertilizer Production

The compositions of ash and slag have been known to contain high mineral compounds with low moisture (Ramzi et al., 2017). Hence, it could be able to mix with waste or organic supplements having high humidity such as sludge from wastewater treatment plants, biodegradable household/ worker waste, and organic waste from processing activities to create a suitable input mixture for fertilizer production.

Suitable mixing will create more symbiotic benefits than only preparing fertilizer from single materials such as sludge or ash. Several benefits could be listed below:

- Moisture balance (ash and slag have a moisture content of 11.2%–22.6%, while sludge and organic waste have a moisture content of up to 80%–90%)
- Mineral balance (mineral supplement for organic waste and mineral control for ash, slag...)

- Energy saving (energy to reduce moisture for sludge and disturbance energy to create porosity of the composting bed)
- Ventilation increase for the compost pile
- Balance the ratio of C/N in the mixture (ash and slag have low nitrogen content, while sludge, some type of garbage, have high nitrogen content)
- Well-controlled pH in the composting process (ash and slag have high pH, while domestic solid waste has low pH)

According to the Ministry of Agriculture and Rural Development, in 2019, Vietnam had 814 companies producing fertilizers, of which 235 produced organic fertilizers. The number of microbiological and organic fertilizer companies in HCMC and neighboring provinces is 164 companies, of which HCMC is 123 companies (MOIT, 2020a). The capacity of fertilizer produced, according to the target of the Ministry of Agriculture and Rural Development, is 12 million tons of organic fertilizer per year. According to the ratio of the number of fertilizer companies in HCMC and neighboring provinces, the amount of fertilizer produced is expected to be no less than 50% of the national output, equivalent to more than 6 million tons/per year. Therefore, if 100% of the ash, slag, and sludge are used as raw materials for producing fertilizer, this volume is still much smaller than the demand.

#### 9.4.2.2 Industrial Symbiosis Solutions for the Production of Leveling Materials

With high inorganic contents and low moisture measured in the case of ash and slag, these waste materials are suitable as a partial alternative to sand for application as a leveling material. When ash, slag, and sludge are appropriately mixed, these final products will achieve several positive effects, specifically as follows:

- Waste sludge has a high moisture content, satisfactory size, suitable to fill the grooves caused by slag, large-sized molds (enhance the effect of surface compaction)
- Moisture balance (ash and slag have low moisture, while sludge has high moisture)
- Improve the color of the leveling mixture (due to slag being dark and lacking in aesthetics when applied)
- Balance mineral and heavy metal components, and ensure to be below the hazardous threshold

The need for ground leveling is huge, especially in HCMC when the general elevation of the whole city has a very high subsidence rate, averaging 2 cm/year (JICA, 2022). Preliminary survey data of the Department of Natural Resources and Environment in 2015 showed that the demand for a few large projects conducted in District 4, District 7, Binh Thanh District, Cu Chi District, Nha Be District, Can Gio District from 2015 to 2025 is about 201,250, 859 $m^3$, equal to average 20 million $m^3$ per year (JICA, 2022). Besides, the urban development speed of HCMC in particular, and the whole country in general is enormous, leading to a significant increase in the demand for leveling. The urbanization rate in 2019 was 36.6% and will increase to 51.2% in 2040 (GSO, 2019).

#### 9.4.2.2 Industrial Symbiosis Solutions for the Production of Concrete

Using materials such as ash, slag, and foundry residue to replace sand, stone, and cement in the concrete production process has been studied and applied for a long time. Depending on the intended use of the concrete, the selection of mixed waste components and the replacement rate of materials will be determined. Some of the benefits when symbiosis waste to produce concrete, specifically as follows:

- Slag casting molds can replace sand and stone at a certain rate, while fly ash can replace cement
- Mixture of slag/casting mold with cement/fly ash and wastewater (e.g. paper mill wastewater with high pollutant concentration) can form the required quality concrete, meeting the standards in construction and especially no pollution emission
- Mixtures have the ability to create specific concrete products that cannot be met by using traditional materials (e.g. heavy/light weight, porosity, absorbency of concrete)
- Slag/ casting molds combine with fly ash and plastic/paper to produce lightweight concrete for construction purposes or produce furniture products (tables, chairs, shelves, and so on)

According to the master plan for the development of building materials in Vietnam,[1] Vietnam's construction sand demand in 2020 is about 130 million tons, most of the current sand consumption demand is still based on natural sand exploitation. Particularly for HCMC, the demand for construction sand in 2020 is 20.38 million $m^3$. Due to the very high demand for sand, while this resource is increasingly scarce, therefore, the price of sand is increasing. The government's policy is to use alternative solutions for sand, such as artificial sand milled from stone, replacement sand from ash and slag waste, and foundry residues (Figure 9.2).

According to the MOIT (2020b), there are 65 concrete manufacturing companies in HCMC and neighboring provinces, of which there are 37 companies in HCMC. These companies can use alternative materials if they meet the requirements for product quality and environmental safety. As for the ash and slag waste components, most waste sources are safe in the concrete production application. Some sources have slightly higher composition than required; However, this can be controlled by reducing the rate of substitution, while considering the application of concrete products to targets far from residential areas, such as making embankments, pouring concrete fences, floors of industrial parks, and so on.

#### 9.4.2.3 Industrial Symbiosis Solutions for the Production of Unburnt Brick/Block Brick

Producing unburnt bricks to replace fired bricks is a useful solution to protect the environment. However, at present, domestic enterprises focus mainly on aggregate concrete brick production lines, while aerated autoclaved concrete (AAC) brick production lines, foam concrete bricks... are mainly imported from abroad. Besides, the rate of using unburnt brick although slightly increased, is still very low and not popular in the community. It is necessary to promote investment in advanced unburnt brick production lines and disseminate the potential of unburnt bricks to the community.

Benefits of industrial symbiosis to produce unburnt bricks:

- Slag replaces sand in the production of unburnt bricks and fly ash replaces cement, limiting resource use and emissions
- Limiting the use of energy in the firing process, limiting pollutant emissions (compared to fired bricks)
- The building using unburnt bricks is lightweight. It meets the specific construction requirements.

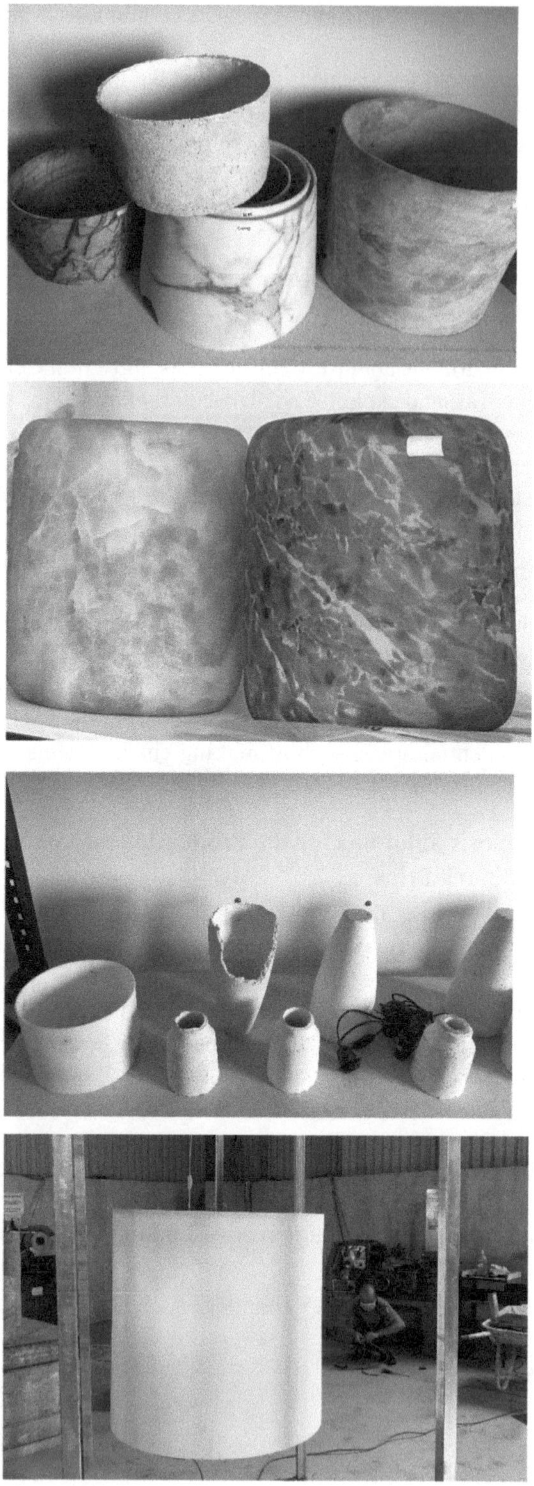

**FIGURE 9.2** Products manufactured from cement, sand, fly ash, bottom ash, paper, and plastic.

## 9.5 INTEGRATED SOLUTIONS FOR VALUE-ADDED PRODUCTS FROM INDUSTRIAL WASTE

To improve the efficiency of environmental management in the direction of a circular economy, some characteristics of the environmental management system in industrial zones should be noted, namely:

- The current system of policies and regulations related to industrial waste management is relatively complete, meeting the environmental protection objectives. However, it still needs practical implementation guidelines to reduce management and disposal costs while enhancing recycling opportunities and industrial symbiosis. There is still a situation in which some policies/regulations have overlapping contents. In contrast, some other policies/regulations have been published, but guidelines have not been issued to be implemented in practice. A specific case of policies to support investment capital and support for the consumption of processed products has been issued. However, the implementation guidance still has many problems and needs more timely and specific. Therefore, it is difficult for small-scale businesses to access these incentives. Specific incentively mechanisms for waste recycling, reuse, and energy recovery need to be improved and consistent. This also causes difficulties in the operation of the business.
- The characteristics of the production facilities are the use of a wide variety of fuels for boilers and thermal furnaces, and various industries. Therefore, ash, slag, and sludge generated from production facilities will differ in composition and characteristics. However, current data on the generation, composition, and characteristics of industrial waste generated from production facilities need to be updated and unsystematic. This is a complex problem for management, as well as difficulty in choosing recycling technology options.
- Waste reuse and recycling enterprises are facing difficulties due to a lack of information on waste sources; there is a lack of scientific basis for implementing waste recycling and reuse technologies, and there are still some legal problems in implementing reuse and recycling.

To improve the efficiency of industrial solid waste management and enhance recycling capacity and industrial symbiosis in HCMC, it is proposed several recommendations, as listed below.

### 9.5.1 MANAGEMENT ASPECTS

- Develop and publish documents guiding waste management activities at the source of generation, collection, treatment, and recycling of waste to relevant parties;
- Overcoming the shortcomings of existing support mechanisms and adding supporting mechanisms related to waste locating, enhancing the efficiency of standardization of waste into raw materials…, to improve the efficiency of socialization of recycling and industrial symbiosis;
- Reviewing the composition and characteristics of waste, completing/adding documents/regulations on hazardous thresholds and application thresholds to produce products from recycled waste to reduce treatment costs and save resources;
- Building and digitizing the database and expanding the opportunity to share the database to improve the management efficiency of the whole system;
- Training human resources, strengthening inspection and supervision, improving the efficiency of using tools/software in inspection and supervision activities, and ensuring the management of the whole life cycle of waste.

### 9.5.2 Technical Aspects

- To meet the goals of improving reuse, recycling, and industrial symbiosis, HCMC needs land planning and the construction of storage and support areas. Industrial waste will be classified and stored in these areas according to recycling characteristics. Besides, research activities can be carried out to propose solutions for reuse, recycling, and industrial symbiosis. The City should consider a separate area for each IZ/EPZ and also a comment area for the whole City, meeting the needs of enterprises inside and outside the IZ/EPZs.
- The government needs to take practical, supportive actions, and at the same time, production facilities must determine their responsibilities in investing in infrastructure, equipments, and improving technology to limit emissions, effectively managing waste at source, and enhancing recycling efficiency.
- Transfer of reuse, recycling, co-treatment/symbiosis technologies that have been successfully applied in many countries around the world such as co-production of cement, concrete production, leveling, fertilizer… At the same time, research and development of industrial symbiosis opportunities with specific characteristics of the locality, each IZ/EPZ and industrial cluster should be expanded.
- Implement in-depth studies on technology and production processes, types and compositions of raw materials, fuels, and so on, to find out the relationship with the quality of recyclable waste to improve recycling efficiency.

## 9.6 SUMMARY AND CONCLUSIONS

This is the first study to evaluate the characteristics of industrial wastes in 17 IZs in HCMC, Vietnam, including ash and slag from different sources and sludge generated from the wastewater treatment plan. The informative results showed that the chemical compositions of each waste met the Vietnam national standard requirement for hazardous waste, which was able to potentially be utilized in various proposed applications. In addition, the present research also indicated that contributors from different sectors might archive effective management of these typical wastes. To further suggestions, in-depth studies should be conducted to thoroughly observe the ultimateness in the case of particular or symbiotic waste.

## ACKNOWLEDGMENT

Sincere thanks to the Department of Science and Technology of HCMC for this research funding and the Management Board of Industrial Parks and Export Processing Zones of HCMC for supporting survey, sampling, and information collection activities.

## NOTE

1 Decision 1469/QĐ- TTg, 22/8/2014 about the master plan on development of building materials in Vietnam to 2020 and vision 2030.

## REFERENCES

Bhatt, A., Priyadarshini, S., Mohanakrishnan, A. A., Abri, A., Sattler, M., & Techapaphawit, S. (2019). Physical, chemical, and geotechnical properties of coal fly ash: A global review. *Case Studies in Construction Materials*, 11, e00263.

DOIT (Department of Industry and Trade in HCMC). (2018). Situation of Industrial Development and Commercial Activities in 2018. DOIT, Vietnam.

Domenech, T., Bleischwitz, R., Doranova, A., Panayotopoulos, D., & Roman, L. (2019). Mapping industrial symbiosis development in Europe_ typologies of networks, characteristics, performance and contribution to the circular economy. *Resources, Conservation and Recycling*, 141, 76–98.

DONRE (Department of Natural Resources and Environment in HCMC). (2020). Planning on Solid Waste Treatment of Ho Chi Minh City in 2025, Vision to 2050. DONRE, Vietnam.

Geissdoerfer, M., Pieroni, M. P., Pigosso, D. C., & Soufani, K. (2020). Circular business models: A review. *J. Clean. Prod.* 277, 123741.

GSO (General Statistics Office of Vietnam). (2019). Statistical Summary Book of Viet Nam 2019. GSO, Vietnam.

HEPZA (Ho Chi Minh City Export Processing and Industrial Zones Authority) (2018). Database management of export processing and industrial zones. HEPZA.

HEPZA (Ho Chi Minh City Export Processing and Industrial Zones Authority). (2020). Statistical Data. HEPZA, Vietnam.

IKEA. (2021). IKEA Sustainability Report FY21. IKEA, Vietnam.

JICA. (2022). Assessing and Implementing Measures Against Land Subsidence *in Ho Chi Minh City (HCMC)*. JICA, Vietnam.

MOIT (Ministry of Industry and Trade). (2020a). The List of Enterprises Licensed to Produce Organic Fertilizers. MOIT.

MOIT (Ministry of Industry and Trade). (2020b). The List of Concrete Manufacturing Companies. MOIT.

MONRE. (2019). Report on current national environmental condition, a focus on solid waste management (in Vietnamese).

Montes-Hernandez, G., Pérez-López, R., Renard, F., Nieto, J. M., & Charlet, L. (2009). Mineral sequestration of CO2 by aqueous carbonation of coal combustion fly-ash. *J. Hazard. Mater.* 161(2), 1347–1354. doi:10.1016/j.jhazmat.2008.04.104

Nihalani, S. A., Mishra, Y. D., & Meeruty, A. R. (2020). Handling and utilisation of fly ash from thermal power plants. In: S. K. Ghosh & V. Kumar (Eds.), *Circular Economy and Fly Ash Management* (pp. 1–11). Singapore: Springer.

Oanh, L. T. K., Hien, H. T. T., Loi, H. T., Khoa, N. D. (2022). Setting up a Database of Industrial Wastes from Industrial Zones in HCMC and Propose Solution Guideline to Treat and Use. Funded by Department of Science and Technology in HCMC, Vietnam.

Ramzi Hannan, N. I. R., Shahidan, S., Ali, N., & Maarof, M. Z. (2017). A comprehensive review on the properties of coal bottom ash in concrete as sound absorption material. *MATEC Web Conf.* 103, 01005.

Salhofer, S., Jandric, A., Soudachanh, S., Le Xuan, T., & Tran, T. D. (2021). Plastic recycling practices in Vietnam and related hazards for health and the environment. *Int. J. Environ. Res. Public Health* 18(8), 4203.

Schneider, P., Anh, L. H., Wagner, J., Reichenbach, J., & Hebner, A. (2017). Solid waste management in Ho Chi Minh City, Vietnam: Moving towards a circular economy? *Sustainability*, 9(2), 286.

# 10 Biomedical Waste Management in COVID-19 Pandemic

*Linh-Thy Le, Xuan-Thanh Bui,*
*Ho Thi Ngoc Ha, Huu Hao Ngo,*
*and Sunita Varjani*

## 10.1 INTRODUCTION

In today's world, it is crucial that we focus on taking care of our communities and ensuring their health and well-being. To do this, we must prioritize the prevention of environmental pollution, specifically the pollution of our soil and water by waste. This is a long-term commitment that will help protect the health of our communities for years to come (Silva et al., 2021). The solid waste (especially biomedical waste) of people containing many pathogens such as dangerous bacteria and viruses (e.g., SARS-CoV-2 virus responsible for the COVID-19 pandemic) is considered a source of maintaining and spreading this agent into the environment and then back to cause disease in humans (Iyer et al., 2021).

In 2020, WHO declared a worldwide human health emergency due to the rapid spread of the COVID-19 pandemic (WHO, 2020). Meanwhile, Rahman et al. (2020) reported more than 5 million deaths annually from unsafe medical waste disposal. This danger does not only impact the health of sanitation workers but also everyone in the community (Mitiku et al., 2022). Several countries have adopted safety measures to manage biomedical waste; however, these measures are insufficient and vary according to the national context. The advantage of proper biomedical waste management is that it can reduce the spread of the SARS-CoV-2 virus (Ganguly and Chakraborty, 2021) and reduce the volume of waste in landfills by reusing specific types of biomedical waste characteristics (e.g., cloth gowns and cloth face masks) (Shammi et al., 2021). The segregation at the initial source and disinfection process of biomedical waste facilitates sustainable management and enables their use for valuable purposes (Ilyas et al., 2020).

This review discusses different biomedical waste management strategies implemented in different countries, the challenges faced in this management process, and possible solutions to overcome these challenges.

## 10.2 TYPES AND SOURCES OF BIOMEDICAL WASTE

It is important to note that biomedical waste is primarily generated from hospitals and veterinary medical premises (BMWM, 2018; Kanwar et al., 2022). This includes a variety of sources, such as hospitals, nursing homes, clinics, laboratories, physicians and veterinarians, dispensaries, and medical transporters. During the COVID-19 pandemic, the definition of biomedical waste has extended to include waste generated during treatment, diagnosis, quarantine, and home care of COVID-19 patients. Any solid waste that has infectiousness factors, such as body fluids and

 DOI: 10.1201/9781003386902-13

secretions of COVID-19 patients, is considered hazardous biomedical waste (Capoor and Parida, 2021). Typically, biomedical waste accounts for 15%–25% of healthcare waste, with the remaining 75%–85% being general waste (Figure 10.1). To segregate the BMW at the source, four color codes are used: yellow, red, white, and blue.

During the COVID-19 outbreak, the widespread use of safety methods to prevent the spread of COVID-19 led to a dramatic increase in BMW emissions. The rapid increase in the number of personal protective equipment (PPE) (e.g., aprons, boots, face shields, gloves, goggles, masks), disinfectants, and other medical equipment, including needles and test kits significantly changed the BMW composition.

## 10.3 BIOMEDICAL WASTE EMISSIONS

Many infectious wastes are produced during the COVID-19 pandemic. The source of contaminated wastes is not only the health facilities including hospitals and medical isolation facility but also households with people infected with COVID-19. The COVID-19 pandemic has resulted in the generation of many infectious wastes. These wastes are not only generated by health facilities such as hospitals and temporary medical facilities but also by households and quarantine centers. To deal with suspected patients of COVID-19, many countries have implemented the common practice of home isolation and quarantine center facilities. Home isolation is used for patients with mild symptoms or travelers, for at least 14 days, so that hospitals can prioritize severe cases requiring intensive medical treatment. This approach is believed to control the COVID-19 outbreak, but it presents new and crucial waste management and treatment challenges. Wastes resulting from households and quarantine facilities with suspected or positive COVID-19 cases may contain viable COVID-19 viruses and possibly be a source of infection. Figure 10.2 summarizes the potential sources of contaminated waste during the COVID-19 pandemic.

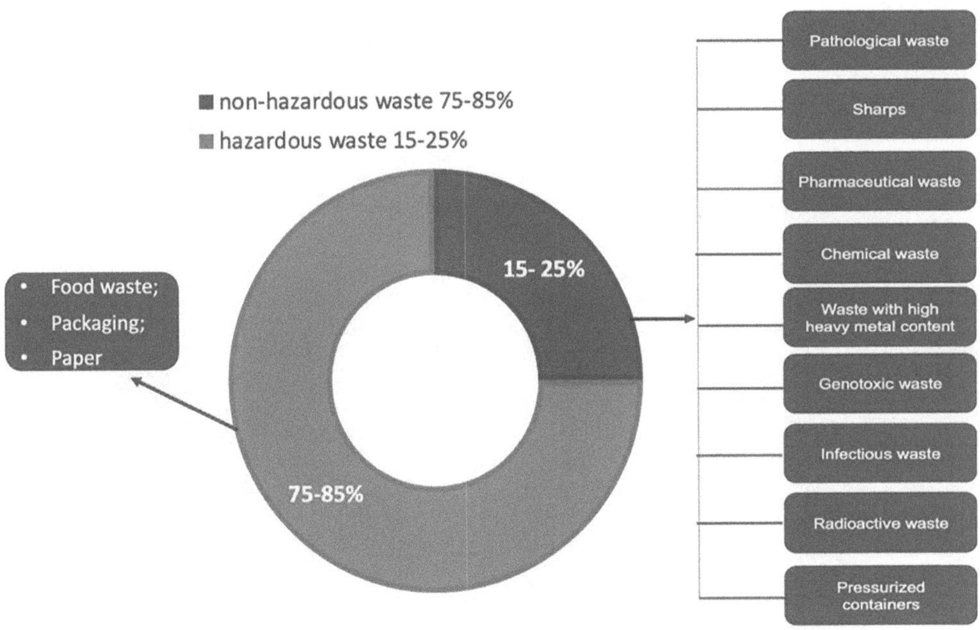


**FIGURE 10.1** Classification of hazardous and nonhazardous waste of biomedical waste.

During a pandemic, all waste generated in healthcare facilities is considered biomedical waste, even though infectious waste only makes up 10%–25% of total healthcare waste (Rupani et al., 2020). Recent studies have shown a rapid increase in infectious waste in hospitals, but the rate of waste generation varies depending on the hospital and location. For example, research in Iran (Kalantary et al., 2021) found that both private and public hospitals generated an average of 110% more biomedical waste during the pandemic, with infectious waste making up an average of 9% more of the total waste composition and a 121% increase compared to pre-pandemic levels. In Wuhan, the daily amount of biomedical waste increased from 40 to 240 tons after the COVID-19 outbreak, while in Ahmadabad, India, the daily amount of biomedical waste increased from approximately 550–1,000 kg with an expected increase up to 3,000 kg/day in containment zones (TOI 2020; Somani et al., 2020). Table 10.1 shows the amount of BMW generated in various places during the pandemic. The significant increase in biomedical waste generation during the pandemic can be attributed to factors such as an increase in hospitalized patients. To reduce the harmful impact of biomedical waste on public health and the environment, it is essential to raise awareness of the issue and find ways to reduce waste generation, including the use of reusable materials that are safe for the environment.

Among biomedical waste, masks are one of the essential components in this category (Sangkham, 2020). Uncontrolled disposal can cause severe environmental impacts (Ganesapillai et al., 2022). Before the pandemic, PPE was used primarily by healthcare workers, but as this pandemic continues, PPE is being widely used by the general population. Almost every country has made using masks mandatory for its citizens (Sarkodie et al., 2020). An online survey result of Badillo-Goicoechea for 38 countries estimated that approximately $14 \times 10^6$ facemasks per day were used in countries during COVID-19 pandemic (Badillo-Goicoechea et al., 2021). As the policy estimates, the total number of disposable masks that are discarded daily in countries such as China, India, the United States, and the United Kingdom accounts for about 4 billion pieces corresponding to the population of those countries (one per person per day) (Zand and Heir, 2020). Asia is forecast to produce the highest number of discarded masks (1.8 billion) per capita per day. Europe follows this with 450 million, Africa with 400 million, Latin America, the Caribbean, and North America with 600 million daily masks.

In addition to the amount of medical waste generated from medical examination and treatment activities, the amount generated from COVID-19 vaccination activities also needs attention. Many

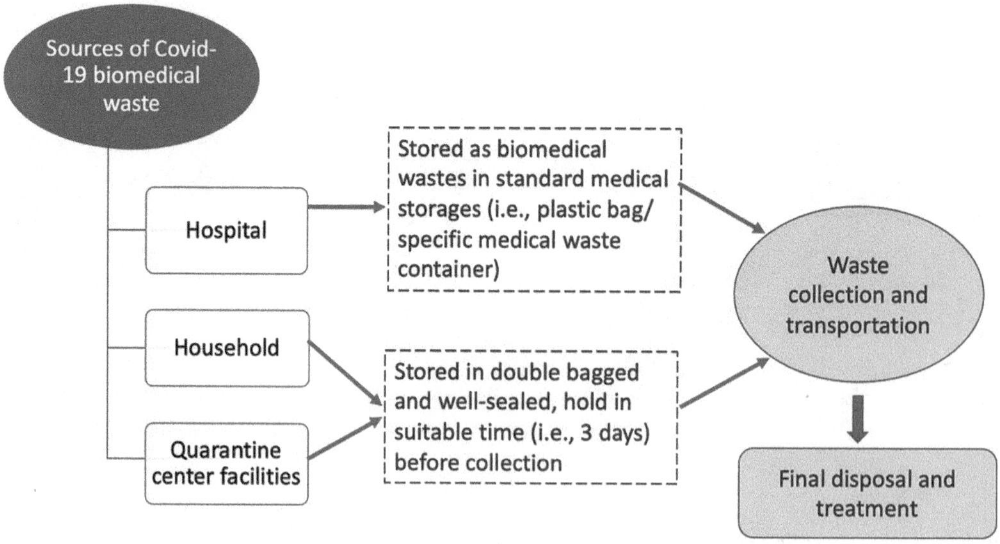


**FIGURE 10.2** Sources of biomedical waste during the COVID-19 pandemic period.

**TABLE 10.1**
**Biomedical Waste Generation in Various Places During the Pandemic**

| Places | Biomedical Waste Generation in Indian States During the COVID-19 Pandemic (kg/day) | References |
|---|---|---|
| Delhi (India) | 11,114 | Kanwar et al. (2022) |
| Haryana (India) | 2,511 | EPCA (2020) |
| Andhra Pradesh | 5,516 | Kanwar et al. (2022) |
| Wuhan | 247,000 | Singh et al. (2020) |
| Catalonia, Spain | 1200,000 | Mondal et al (2022) |
| Manila | 280,000 | Statista (2023) |
| Jakarta | 212,000 | Statista (2023) |
| Bangkok | 160,000 | Statista (2023) |
| Hanoi | 154,000 | Statista (2023) |

organizations around the world are working tirelessly to develop drugs and vaccines to prevent the spread of this virus. While vaccination is necessary to reduce the number of patients with COVID-19, it's important to note that it also generates a large amount of medical waste such as needles, syringes, gloves, and vaccine containers. According to statistics from Our World in Data (Mathieu et al., 2021), between June 2021 and December 2021, about 25–40 million doses of COVID-19 vaccine were administered every day. This means that significant amounts of medical waste are being generated on a daily basis. In addition to vaccination activities, the daily medical examination and treatment of medical staff, the increase of hospitalized patients, and the regulation of wearing masks in daily communication have all contributed to the overall increase of medical waste during the COVID-19 pandemic.

## 10.4 ENVIRONMENTAL SUSTAINABILITY AND TRANSMISSIBILITY OF SARS-COV-2 VIRUS

SARS-CoV-2 can be transmitted through daily activities such as sneezing, coughing, close contact with COVID-19 patients, or direct contact with surfaces contaminated with the SARS-CoV-2 virus (Dhand and Li, 2020). Virus persistence depends on several factors, such as the surface type, the presence and type of organic substrate in which the virus is suspended at the time of surface deposition and drying, and the contact duration. SARS-CoV-2 showed variable persistence on different porous surfaces, such as paper, cardboard, wood, cloth, and mask (Azuma et al., 2020) (Figure 10.3).

Various studies have demonstrated that SARS-CoV-2 is infectious on cloth and wood for 1 day (22°C), about 7 days on a surgical mask (outer layer) (22°C), and 14 days on an N95 mask (20°C) (Chin et al., 2020; Moriyama et al., 2020; Zhang et al., 2020). On the contrary, the SARS-CoV-2 virus could be recovered from a surgical mask after 7 days at ambient temperature and relative humidity (RH) of 65%. SARS-CoV-2 decayed more rapidly when either the humidity or the temperature was increased. SARS-CoV-2 virus could survive up to a month under a cold temperature and low RH conditions of approximately three degrees and 19% (Viana Martins et al., 2022). The long persistence of the COVID-19 virus could increase the risk of community transmission. Therefore, it is considered that proper medical waste management can reduce the spread of the COVID-19 virus (Nzediegwu and Chang, 2020; Ramteke and Sahu, 2020; Sharma et al., 2020; Yang et al., 2021). Some studies have shown that SARS-CoV-2 is quite sensitive to high temperatures. Virus

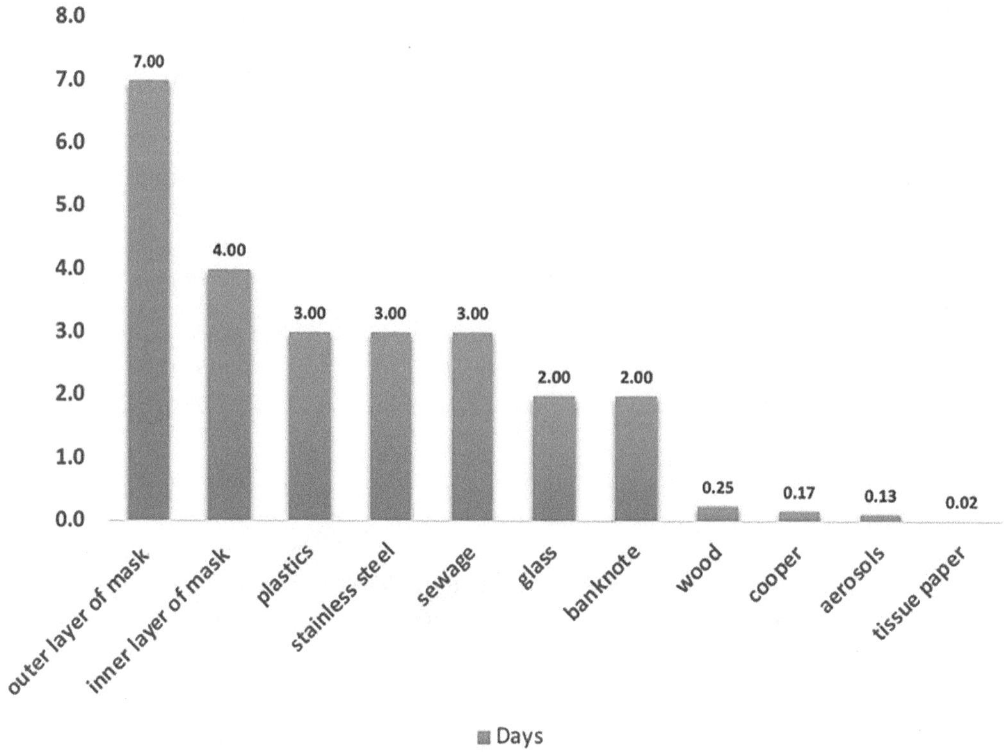


**FIGURE 10.3** Transmissibility time of SARS-CoV-2 virus.

survival decreases with an increase in temperature. Maintaining temperatures above 60°C for more than 60min usually inactivates most viruses [40]. SARS-CoV-2 can be destroyed at temperatures between 40°C and 60°C. Besides, SARS-CoV-2 can be inactivated by standard disinfection procedures (Saadatpour and Mohammadipanah, 2020). The results of Chin et al. (2020) indicated that infectious SARS-CoV-2 could not be detected after incubation with soap for 5min at room temperature. In short, the data on the survival time of the COVID-19 virus on different materials and environmental conditions is important information for developing suitable biomedical waste management and treatment methods during the COVID-19 pandemic.

## 10.5 WASTE MANAGEMENT STRATEGIES DURING THE COVID-19 PANDEMIC

It is a matter of grave concern that the amount of medical waste is on the rise due to the growing medical needs of people and the number of patients in hospitals. The amount of waste generated by hospitals is now two to four times higher than usual, owing to the surge in usage of costumes, masks, gloves, equipment, needles, syringes, fluid lines, and medicines. Even the usage of masks is contributing to the waste, as they are being used frequently and then disposed of. Isolation areas with thousands of people are further contributing to the waste with the use of masks, protective clothing, and ready-made food. The use of disinfectant chemicals like Chloramine B by health authorities is also adding to the toxicity of the environment. Burning medical waste is a common method used in hospitals, isolation areas, and urban waste treatment areas, which is leading to air pollution. Consequently, the increase in medical waste will have a profound effect on sustainable waste management activities in the future (Das et al., 2021). Good management of medical waste will be essential to avoid the potential transmission of pathogens from the waste mass to staff and the community, especially during the COVID-19 pandemic (Kalantary et al., 2021). Managing biomedical waste is a crucial aspect of healthcare, and an efficient biomedical waste management strategy should address several main contents. These include compliance with regulations, the responsibilities of healthcare workers and the community, the classification of healthcare waste from sources, specific treatment procedures for biomedical waste, and training programs about biomedical waste management. It is important to ensure that all of these aspects are properly addressed to minimize the potential harm that biomedical waste can cause (Datta et al., 2018; Manzoor and Sharma, 2019; Das et al., 2021). Various countries have applied strategies based on their capacity and resources to manage the vast and infectious biomedical waste produced during the COVID-19 pandemic.

According to the European Union (EU) guidelines, biomedical waste generated during the COVID-19 pandemic should be considered infectious waste, and the capacity to manage this waste at medical facilities needs to be strengthened. Facilities' conditions (e.g., large enough storage area, adequate and suitable facilities) should be adequately prepared in case biomedical wastes need to be stored for extended periods before incineration or disposal for unexpected reasons. Containers made of rigid and sealed material used to keep waste should be in separate areas where unauthorized persons cannot enter. Most importantly, the exterior surfaces and the entire biomedical waste storage area should be sprayed with disinfectants to avoid virus transmission (Das et al., 2021). Table 10.2 presents the ways suggested by the WHO to segregate biomedical waste. Besides, all healthcare workers and sanitation workers in the area should follow appropriate safety measures.

During the COVID-19 pandemic, the amount of medical waste generated in Wuhan far exceeded the city's existing capacity for treatment, approximately six times more than before the pandemic's peak (Singh et al., 2020). At its peak, approximately 247 tons of medical waste was generated per day, compared to an average of about 50 tons per day before the outbreak (Wei, 2020). This led to large amounts of waste being stored for several days in medical facilities, waiting to be transported for disposal (Nzediegwu and Chang, 2020). To prevent waste from piling up during emergencies like COVID-19, it is important to have equipment ready for medical waste treatment facilities. Medical waste treatment facilities have become more automated using Internet of Things (IoT) technology with minimal staffing, especially in developing countries where the capacity of medical waste treatment facilities is limited (Chen et al., 2020).

In Hubei (China), BMW has been separated by waste treatment staff in the hospital at the place of generation and packaging. A 0.5% chlorine solution is used to disinfect waste. The two-layer

**TABLE 10.2**
**WHO-Recommended Segregation Scheme (WHO, 2014)**

| Waste Types | Type of Container | Color of Container |
|---|---|---|
| Infectious | Leak-proof strong plastic bag placed in a container (bag for highly infectious waste should be capable of being autoclaved) | Yellow with biohazard symbol (highly infectious waste should be marked as HIGHLY INFECTIOUS) |
| Sharp | Puncture-proof container | Yellow and marked SHARP with biohazard symbol |
| Pathological | Leak-proof strong plastic bag placed in a container | Yellow with biohazard symbol |
| Chemical and pharmaceutical | Plastic container | Brown labeled with appropriate hazard symbol |
| Radioactive | Lead box | Labeled with radiation symbol |
| General | Plastic bag inside a container which is disinfected after use | Black label |

packaging process is applied to the entire BMW before being placed in temporary storage in the hospital. In addition, some hospitals may use an autoclave to disinfect before transferring waste to a licensed landfill. In places with difficult economic conditions, some hospitals have applied on-site incineration to treat medical waste. Portable incineration systems or autoclaves were provided to support the additional medical waste generated during the outbreak. Only licensed transport vehicles may be used to transport BMW. The BMW parking area is also disinfected and separated from the driver to avoid the risk of infection (ADB, 2020).

Several principles have been applied in the management of BMW during the pandemic in Jordan. Principles include reducing unnecessary medical waste, separating ordinary waste from hazardous waste, and using appropriate BMW handling to minimize risks to medical personnel and the surrounding environment. If medical waste disposal personnel are infected with COVID-19, they must use ultrafiltration masks, long-sleeved protective clothing, caps, shoes, elastic leather gloves, and goggles. Authorities closely monitor the work to ensure that local rules and regulations regarding cleaning and disinfection are strictly followed. Waste originating from COVID-19 patients is quickly disposed of daily. Temporary and permanent storage areas, containers, and bags for medical waste are disinfected to prevent the spread of the virus (ISWA-Jordan, 2020).

In Vietnam, a developing country, medical waste management during the COVID-19 pandemic was done according to the Vietnam Ministry of Health Regulations. Medical waste such as face masks, towels, tissue, and PPE generated from isolation areas at complete medical isolation facilities and establishments must be immediately segregated into infectious waste containers. Infectious waste containers are regulated in yellow color with lids, lined with bags, with a warning symbol for waste containing pathogenic substances. A note "Waste with risk of containing SARS-CoV-2" was applied outside the container. If there is a case or suspected case of COVID-19, all waste generated in this person's isolation room must be collected and treated as infectious waste with the risk of containing SARS-CoV-2. Other wastes shall be collected and treated according to the Joint Circular 58/2015/TTLT-BYT-BTNMT (Medical waste management regulation) (Nguyen et al., 2021). Before collection, the waste bag is tied tightly at the mouth of the bag and continued to be placed in the second infectious waste bag, sealed, and then placed in the infectious waste collection bin. Separately collect the waste container containing SARS-CoV-2 from the isolation area/room to the temporary waste storage area in the facility at least twice a day or as needed. During the collection process, the waste container must have a tight lid to ensure the waste does not fall or leak out. The infectious waste collection area should be located at the end of the wind direction, convenient in the transportation route, must have a roof, and the ground must not be flooded. Containers and utensils containing potentially hazardous waste containing SARS-CoV-2 must be cleaned and disinfected immediately with a disinfectant solution containing 0.5% active chlorine. Carry out transporting the waste container with the risk of containing SARS-CoV-2 to the nearest centralized hazardous waste treatment facility or hazardous waste treatment facility with biomedical waste treatment function by specific vehicle for handling during the day. During transportation, the waste container is always covered with a tight lid to ensure that it does not drop or leak waste. All surfaces in isolation areas and clinics for people with or suspected of COVID-19, including covers of healthcare equipment and vehicles, need daily surface cleaning for environmental disinfection. Chemicals for cleaning and disinfecting are mixed according to regulations containing 0.1% active chlorine. Besides, persons performing cleaning and disinfection of environmental surfaces must be instructed to be fully equipped with PPE.

From the above examples, it can be seen that different countries have taken different measures to manage medical waste properly. Adequate safety measures and working strategies can enable proper medical waste management without spreading the virus to others. Disinfection followed by proper segregation and on-site disposal of BMW can also lead to better and more safe management of medical waste in the COVID-19 pandemic.

## 10.6 CHALLENGES

The management of medical waste has been a challenging task, especially in developing countries. The COVID-19 pandemic has made this issue more critical. The surge in demand for masks, PPE, and disposable plastic items has become a global concern for governments. This problem is even more severe in countries with weak economies, where biomedical waste is often disposed of with regular waste in landfills. This poses a significant risk to public health since scavengers gather recyclable materials without proper protective gear. It is crucial to find effective solutions to address this problem and promote public health (El-Ramady et al., 2021). Decontamination, inadequate treatment, and improper disposal of disposable biomedical waste in quarantine and healthcare facilities can be another potential source of virus transmission. Various toxic pollutants and pathogens can survive and grow on medical waste surfaces, posing harmful risks to human health and the environment (Padmanabhan and Barik, 2019). Besides, during the COVID-19 pandemic, one patient can generate an average of about 3.5 kg/day of medical waste, including facial masks, plastic gloves, and protective clothes. These PPEs lead to a gradually increased quantity of plastic waste (Das et al., 2021). If each person simply throws away one used mask during this time, and only 1% of those masks are not disposed of properly, the world's environment could be contaminated with an enormous amount of plastic waste. Not only that, but there's also a significant risk of infection with the SARS-CoV-2 virus in the environment. It's important to dispose of medical waste properly and do our part to keep our surroundings clean and safe.

We are facing some serious hurdles in managing biomedical waste during the pandemic. The surge in medical and infectious waste from hospitals and households has created an unprecedented challenge. Many are worried about the capacity of existing treatment facilities, which have not been able to keep up with the increase in BMW. In addition, the generation of a wider variety of disposable wastes is also a concern. Lastly, there is a high risk of virus transmission due to infectious waste in areas with underdeveloped economies. These challenges require careful consideration and action to ensure the safe and responsible management of biomedical waste.

## 10.7 RECOMMENDATIONS AND PERSPECTIVES

Waste management is essential to prevent undesirable effects on human health and the environment because of the disposal of drug-resistant pathogens, hazardous biological and chemical agents from biomedical waste, especially COVID-19 biomedical waste, into the environment. Numerous studies have shown that following adequate safety measures in the collection, treatment, and disposal of biomedical waste is essential to dispose of this high-risk infectious waste. First of all, the distinct separation and collection of biomedical waste (or COVID-19 waste) should be carried out at generated places (e.g., hospital, isolation area, or house of a person infected with COVID-19) to prevent waste collectors from opening sacks. These hazardous wastes must be contained in containers (e.g., bags, bins) with double seals, the used volume must be 2/3 of the container's maximum volume, and must be labeled with classification. It is important to remember that COVID-19 medical waste should not be disposed of by mixing it with domestic solid waste (Luhar et al., 2022). In addition, vehicles participating in COVID-19 waste collection must be disinfected by spraying 1% NaOCl solution after each waste collection. Immediately after removing PPE and masks, workers should avoid touching their face, nose, mouth, and eyes; they should do so only after using 70% alcohol sanitizer (Behera, 2021).

Furthermore, during a pandemic, consumables such as surgical masks and gloves (PPE) are used frequently, leading to an increase in biomedical waste emissions. Due to the small size and light weight of PPEs, these wastes can be disposed of with solid waste. Therefore, handling highly infectious wastes needs to be carefully monitored because their infectivity is still relatively high in the first 7 days. Biomedical hazardous wastes should be stored for longer than 7 days to ensure safety.

Besides, it's crucial to prioritize the reduction of biomedical waste generation by exploring reuse measures. Through further study, we can discover ways to recycle and reuse materials, which will effectively limit the spread of harmful microorganisms. Moreover, recycling technology can provide easier access to personal protective measures as they can be reused instead of being constantly disposed of. By giving more attention to reuse measures, we can make a significant positive impact on both public health and the environment. Autoclaving can increase the amount of recyclable material, which may help reduce landfill waste volume during the COVID-19 pandemic (Zand and Heir, 2020). An autoclave utilizes the basic principle of steam sterilization at high pressure (higher than 2.2 atmospheres) and temperature (above 132°C) in an atmosphere 100% filled with saturated water vapor (Kenny and Priyadarshini, 2021). Viruses and bacteria are exposed to direct steam at the required temperature and pressure for the defined time. Therefore, autoclaving of healthcare solid waste can improve the efficiency of the recycling process as the process can effectively disinfect the contaminated waste. For example, right after using the injection equipment, the needle is separated from the plastic part with a needle cutter/needle destructor, and then the needle will fall into a container. The plastic part of the syringe is dumped into a container for recycling after autoclaving. Along with the advantage, autoclaving treatment has some disadvantages, including being unsuitable for large-scale treatment of BMW, high investment price, and only reducing the mass but no weight (UNDP, 2022).

It is imperative to keep in mind that human coronaviruses can remain infectious for up to 9 days on nonliving surfaces. To combat this, it has been shown that surface disinfection using 0.1% sodium hypochlorite, 0.5% hydrogen peroxide, or 62%–71% ethanol is effective in just 1 min against coronaviruses (Kampf, 2020). It is also believed that this approach will work against the SARS-CoV-2 virus. It is important to take measures to guarantee that surfaces are appropriately disinfected to prevent the virus from spreading. Therefore, to decrease the risk of exposure and ensure the safety of collection workers and medical waste treatment, it is beneficial to use disinfectants and store medical waste for more than 7 days in proper storage. This will aid in minimizing the spread of pathogenic microorganisms and keeping everyone safe.

During these challenging times of the pandemic, it is essential to explore methods to enhance the use of medical waste. A promising approach is to use pyrolysis and incineration to create value-added products. For example, incinerator ash from medical waste can be reused to make Portland cement (Matalkah, 2023). Moreover, organic waste can be transformed into compost and bioenergy (Zand and Heir, 2020). By discovering innovative ways to repurpose medical waste, we can contribute to a circular economy and efficiently manage our resources.

Regular training on the impact of hazardous waste on public health and the environment is vital for city health and sanitation workers. Proper management of hazardous or infectious medical waste can have a positive effect on the community's safety and well-being. It is essential for the government, local agencies, and waste treatment facilities to create a program that promotes awareness about the safe handling of COVID-19 waste. By working together to promote safe practices, we can safeguard ourselves and our environment.

## 10.8 SUMMARY AND CONCLUSIONS

Managing biomedical waste has proven to be a significant challenge worldwide, especially developing countries, leading to an increase in solid waste. However, the primary issue is the lack of advanced hazardous medical waste management systems. Fortunately, there are several viable solutions to address this problem, including waste reduction at the source, color labeling for waste segregation, and PPE reuse. Incineration is a highly effective option that can generate electrical energy and reduce landfill waste by up to 85%. Nevertheless, it's crucial to consider both the environmental and human health implications, in addition to reducing waste production. Adequate disinfection measures and PPE are critical, and healthcare providers and waste collectors must receive comprehensive training in these areas. By adopting safe practices, we can minimize the risk of infection and protect ourselves and the environment.

## ACKNOWLEDGMENT

We acknowledge Ho Chi Minh City University of Technology (HCMUT), VNU-HCM for supporting this study.

## REFERENCES

ADB, 2020. Managing Infectious Medical Waste during the COVID-19 Pandemic. https://www.adb.org/sites/default/files/publication/578771/managing-medical-waste-covid19.pdf

Azuma, K., Yanagi, U., Kagi, N. et al., 2020. Environmental factors involved in SARS-CoV-2 transmission: Effect and role of indoor environmental quality in the strategy for COVID-19 infection control. *Environmental Health and Preventive Medicine* 25, 66.

Badillo-Goicoechea, E., Chang, T.H., Kim, E., LaRocca, S., Morris, K., Deng, X., Chiu, S., Bradford, A., Garcia, A., Kern, C., Cobb, C., Kreuter, F., Stuart, E.A., 2021. Global trends and predictors of face mask usage during the COVID-19 pandemic. *BMC Public Health* 21(1):2099. doi: 10.1186/s12889-021-12175-9.

Behera, B.C., 2021. Challenges in handling COVID-19 waste and its management mechanism: A review. *Environmental Nanotechnology, Monitoring and Management* 15, 100432. doi:10.1016/j.enmm.2021.100432.

BMWM, 2018. Biomedical Waste Management Rules. Ministry of Environment, Forest and Climate Change, Delhi, India.

Capoor, M.R., Parida, A., 2021. Current perspectives of biomedical waste management in context of COVID-19. *Indian Journal of Medical Microbiology* 39(2), 171–178. doi:10.1016/j.ijmmb.2021.03.003.

Chen, Y., and Guo, C., 2020. Handbook of emergency disposal and management of medical waste in china. http://bcrc.tsinghua.edu.cn/col/1256347643949/2020/06/22/1592834394326.html.

Chin, A.W.H., Chu, J.T.S., Perera, M.R.A., Hui, K.P.Y., Yen, H.-L., Chan, M.C.W., Peiris, M., Poon, L.L.M., 2020. Stability of SARS-CoV-2 in different environmental conditions. *Lancet* 1(1), e10. doi:10.1016/S2666-5247(20)30003-3.

Das, A.K., Islam, M.N., Billah, M.M., Sarker, A., 2021. COVID-19 pandemic and healthcare solid waste management strategy: A mini-review. *Science of the Total Environment* 778, 146220. doi:10.1016/j.scitotenv.2021.146220

Datta, P., Mohi, G., Chander, J., 2018. Biomedical waste management in India: Critical appraisal. *Journal of Laboratory Physicians* 10(1), 6. doi:10.4103/jlp.jlp_89_17

Dhand, R., Li, J., 2020. Coughs and sneezes: Their role in transmission of respiratory viral infections, including SARS-CoV-2. *American Journal of Respiratory and Critical Care Medicine*. doi:10.1164/rccm.202004-1263pp

El-Ramady, H., Brevik, E.C., Elbasiouny, H., Elbehiry, F., Amer, M., Elsakhawy, T., Omara, A.E., Mosa, A.A., El-Ghamry, A.M., Abdalla, N., Rezes, S., Elboraey, M., Ezzat, A., Eid, Y., 2021. Planning for disposal of COVID-19 pandemic wastes in developing countries: A review of current challenges. *Environmental Monitoring and Assessment* 193(9), 592. doi:10.1007/s10661-021-09350-1.

EPCA (2020) The Environment Pollution (Prevention and Control) Authority (EPCA).https://www.epca.org.in/EPCA-Reports1999-1917/EPCA-reportno-112.pdf. Accessed on 25 Feb 2024.

Ganesapillai, M., Mondal, B., Sarkar, I., Sinha, A., Ray, S.S., Kwon, Y-N., Nakamura, K., Govardhan, K., 2022. The face behind the COVID-19 mask: A comprehensive review. *Environmental Technology & Innovation* 28, 102837. doi:1016/j.eti.2022.102837.

Ganguly, R.K., Chakraborty, S.K., 2021. Integrated approach in municipal solid waste management in COVID-19 pandemic: Perspectives of a developing country like India in a global scenario. *Case Studies in Chemical and Environmental Engineering* 3, 100087. doi:10.1016/j.cscee.2021.100087

Ilyas, S., Srivastava, R.R., Kim, H., 2020. Disinfection technology and strategies for COVID-19 hospital and bio-medical waste management. *Science of the Total Environment* 141652. doi:10.1016/j.scitotenv.2020.141652

ISWA-Jordan, 2020. Country Specific Waste Management Responses, COVID-19 Response International Knowledge Sharing on Waste Management. https://www.iswa.org/fileadmin/galleries/0001_COVID/Jordan_Solid_Waste_Handling_Manual_Coronavirus_crisis-ENG.pdf

Iyer, M., Tiwari, S., Renu, K., Pasha, M.Y., Pandit, S., Singh, B., Raj, N., Krothapalli, S., Kwak, H.J., Balasubramanian, V., Jang, S.B., Dileep Kumar, G., Uttpal, A., Narayanasamy, A., Kinoshita, M., Subramaniam, M.D., Nachimuthu, S.K., Roy, A., Valsala, Gopalakrishnan, A., Ramakrishnan, P., Cho, S.G., Vellingiri, B., 2021. Environmental survival of SARS-CoV-2: A solid waste perspective. *Environmental Research* 197, 111015. doi:10.1016/j.envres.2021.111015

Kalantary, R.R., Jamshidi, A., Mofrad, M.M.G., Jafari, A.J., Heidari, N., Fallahizadeh, S., Hesami, Arani M., Torkashvand, J., 2021. Effect of COVID-19 pandemic on medical waste management: A case study. *Journal of Environmental Health Science & Engineering* 19(1), 831–836. doi:10.1007/s40201-021-00650-9.

Kampf, G., 2020. Potential role of inanimate surfaces for the spread of coronaviruses and their inactivation with disinfectant agents. *Infection Prevention in Practice* 2(2), 100044. doi:10.1016/j.infpip.2020.100044

Kanwar, V.S., Sharma, A., Rinku, M.K., Srivastav, A.L., Soni D.K., 2022. An overview for biomedical waste management during pandemic like COVID-19. *International Journal of Environmental Science and Technology*. doi:1007/s13762-022-04287-5.

Kenny, C., Priyadarshini, A., 2021. Review of current healthcare waste management methods and their effect on global health. *Healthcare* 9, 284. doi:10.3390/healthcare9030284

Luhar, I., Luhar, S., Abdullah, M.M.A.B., 2022. Challenges and impacts of COVID-19 pandemic on global waste management systems: A review. *Journal of Composites Science* 6, 271.

Manzoor, J., Sharma, M., 2019. Impact of biomedical waste on environment and human health. *Environmental Claims Journal* 1–24. doi:10.1080/10406026.2019.1619265

Matalkah, F., 2023. Recycling of hazardous medical waste ash toward cleaner utilization in concrete mixtures. *Journal of Cleaner Production* 400, 136736, doi:1016/j.jclepro.2023.136736.

Mathieu, E., Ritchie, H., Ortiz-Ospina, E., Roser, M., Hasell, J., Appel, C., Giattino, C., Rodés-Guirao, L., 2021. A global database of COVID-19 vaccinations. *Nature Human Behaviour* 5(7), 947–953. doi:10.1038/s41562-021-01122-8.

Mitiku, G., Admasie, A., Birara, A., Yalew, W., 2022. Biomedical waste management practices and associated factors among health care workers in the era of the COVID-19 pandemic at metropolitan city private hospitals, Amhara region, Ethiopia, 2020. *PLOS ONE* 17(4), e0266037.

Mondal, R., Mishra, S., Pillai, J.S., Sahoo, M.C., 2022. COVID 19 Pandemic and biomedical waste management practices in healthcare system. *Journal of Family Medicine and Primary Care* 11, 439–446.

Moriyama, M., Hugentobler, W.J., Iwasaki, A., 2020. Seasonality of respiratory viral infections. *Annual Review of Virology* 7(1), 83–101. doi:10.1146/annurev-virology-012420-022445.

Nguyen, T.D., Kawai, K., Nakakubo, T., 2021. Estimation of COVID-19 waste generation and composition in Vietnam for pandemic management. *Waste Management & Research*, 39(11), 1356–1364. doi:10.1177/0734242X211052849

Nzediegwu, C., Chang, S.X., 2020. Improper solid waste management increases potential for COVID-19 spread in developing countries. *Resources, Conservation and Recycling* 161, 104947. doi:10.1016/j.resconrec.2020.104947.

Padmanabhan, K.K., Barik, D., 2019. Health hazards of medical waste and its disposal. *Energy from Toxic Organic Waste for Heat and Power Generation* 2019, 99–118. doi:10.1016/B978-0-08-102528-4.00008-0.

Rahman, M.M., Bodrud-Doza, M., Griffiths, M.D., Mamun, M.A., 2020. Biomedical waste amid COVID-19: Perspectives from Bangladesh. *Lancet Global Health* 8(10), e1262. doi:10.1016/S2214-109X(20)30349-1.

Ramteke, S., Sahu, B.L., 2020. Novel coronavirus disease 2019 (COVID-19) pandemic: Considerations for the biomedical waste sector in India. *Case Studies in Chemical and Environmental Engineering* 2, 100029. doi:10.1016/j.cscee.2020.100029.

Rupani, P.F., Nilashi, M., Abumalloh, R.A., Asadi, S., Samad, S., Wang. S., 2020. Coronavirus pandemic (COVID-19) and its natural environmental impacts. *International Journal of Environmental Science and Technology*, 17, 4655–4666

Sarkodie, S.A., Owusu, P.A., 2020. Impact of COVID-19 pandemic on waste management. Environmental Development & Sustainability, 1–10.

Saadatpour, F., Mohammadipanah, F., 2020. Physicochemical susceptibility of SARS-CoV-2 to disinfection and physical approach of prophylaxis. *Health Science Reports* 3(4), e213. doi:10.1002/hsr2.213.

Sangkham, S., 2020. Face mask and medical waste disposal during the novel COVID-19 pandemic in Asia. *Case Studies in Chemical and Environmental Engineering* 2, 100052. doi:10.1016/j.cscee.2020.100052.

Shammi, M., Behal, A., Tareq, S.M., 2021. The escalating biomedical waste management to control the environmental transmission of COVID-19 pandemic: A perspective from two south Asian countries. *Environmental Science & Technology* 55(7), 4087–4093. doi:10.1021/acs.est.0c05117

Sharma, H. B., Vanapalli, K. R., Cheela, V. S., Ranjan, V. P., Jaglan, A. K., Dubey, B., Goel, S., & Bhattacharya, J., 2020. Challenges, opportunities, and innovations for effective solid waste management during and post COVID-19 pandemic. Resources, Conservation and Recycling, 162, 105052. https://doi.org/10.1016/j.resconrec.2020.105052

Silva, L.F.O., Oliveira, M.L.S., Crissien, T.J., Santosh, M., Bolivar, J., Shao, L., Schindler, M., 2021. A review on the environmental impact of phosphogypsum and potential health impacts through the release of nanoparticles. *Chemosphere* 131513. doi:10.1016/j.chemosphere.2021.131513

Singh, N., Tang, Y., Zhang, Z., Zheng, C., 2020. COVID-19 waste management: Effective and successful measures in Wuhan, China. *Resources, Conservation and Recycling* 163, 105071. doi:10.1016/j.resconrec.2020.105071.

Somani, M., Srivastava, A.N., Gummadivalli, S.K., Sharma, A. 2020. Indirect implications of COVID-19 towards sustainable environment: an investigation in Indian context. Bioresour Technol Rep. 11:100491. doi: 10.1016/j.biteb.2020.100491.

Statista, 2023. Daily Healthcare Waste Generation Before and During the COVID-19 Pandemic in Asia, by Select City. https://www.statista.com/statistics/1167512/healthcare-waste-generation-before-during-COVID-19-asia-by-city/#statisticContainer (accessed on 3, 2023)

Times of India (TOI) (2020) 350 Tonnes of trash disappear from streets of Ahmedabad. http://timesofindia.indiatimes.com/articleshow/75039037.cms?utm_source=contentofinterest&utm_medium=text&utm_campaign=cppst

UNDP, 2022. Evaluation of the Medical Waste Management System in the Kyrgyz Republic. https://www.undp.org/kyrgyzstan/publications/evaluation-medical-waste-management-system-kyrgyz-republic (accessed on 4, 2023).

Viana Martins, C.P., Xavier, C.S.F., Cobrado, L., 2022. Disinfection methods against SARS-CoV-2: A systematic review. *Journal of Hospital Infection* 119, 84–117. doi:10.1016/j.jhin.2021.07.014.

Wei, G., 2020. Medical waste management experience and lessons in covid-19 outbreak in Wuhan. https://www.waste360.com/medical-waste/medical-waste-management-experience-and-lessons-covid-19-outbreak-wuhan.

WHO, 2014. Segregation, storage, and transport of health care waste. In Chartier, Y., Emmanuel, J., Pieper, U., Pruss, A., Rushbrook, P., Stringer, R., Townend, W., Wiburn, S. and Zghondi, R. Eds. *Safe Management of Wastes Fromhealth Care Activities*, 2nd edn. World Health Organisation, Geneva, pp. 77–104. https://apps.who.int/iris/bitstream/10665/85349/1/9789241548564_eng.pdf?ua=1 (accessed November 13, 2022).

WHO, 2020. Novel Coronavirus *(2019-nCoV) Situation Report 11*. https://www.who.int/docs/default-source/coronaviruse/situation-reports/20200131-sitrep-11-ncov.pdf?sfvrsn=de7c0f7_4 (accessed February 18, 2023)

Yang, L., Yu, X., Wu, X., Wang, J., Yan, X., Jiang, S., Chen, Z., 2021. Emergency response to the explosive growth of health care wastes during COVID-19 pandemic in Wuhan, China. Resources, Conservation and Recycling, 164, 105074. https://doi.org/10.1016/j.resconrec.2020.105074

Zand, A.D., Heir, A.V., 2020. Emerging challenges in urban waste management in Tehran, Iran during the COVID-19 pandemic. *Resources, Conservation and Recycling* 162, 105051. doi:10.1016/j.resconrec.2020.105051

Zhang, W., Du, R.H., Li, B., Zheng, X.S., Yang, X.L., Hu, B., Wang, Y., Xiao, G-F., Yan, B., Shi, Z-L., Zhou, B., 2020. Molecular and serological investigation of 2019-nCoV infected patients: Implications of multiple shedding routes. *Emerging Microbes & Infections* 9, 386–389. doi:10.1080/22221751.2020.1729071

# Index

FSC
www.fsc.org
MIX
Papier | Fördert
gute Waldnutzung
FSC® C083411

Kolibri 360 GmbH
Ferdinand-Jühlke-Straße 7
99095 Erfurt, Deutschland
produktsicherheit@kolibri360.de